AN INTRODUCTION TO CIVIL ENGINEERING

Edited by

June Blyth Orillana Royol

Delve Publishing

An Introduction to Civil Engineering

Edited by June Blyth Orillana Royol

Edition 2018

Delve Publishing

Published by

Delve Publishing LLC

708 3rd Avenue, 6th Floor

New York

NY 10801

United States of America

www.delvepublishing.com

ISBN: 978-1-68095-679-5 (Hardback)

Printed in United States of America

Library of Congress Control Number: 2016961988

About the Editor

June Blyth Orillana Royol

June Blyth Royol completed her study at Negros Oriental State University. An accredited Contractors' and Consultants' Materials Engineer I since April 2015. Published a book entitled "Optimum Water-Cement Ratio for 50% Recycled Concrete as Coarse Aggregate for Building Construction". She is currently working as a Licensed Civil Engineer at the Department of Public Works and Highways.

Table of Contents

3 STRUCTURAL ENGINEERING: ELEMENTS AND SYSTEMS

4 TRANSPORTATION ENGINEERING

5 ROAD SURFACE ENGINEERING

6 ENVIRONMENTAL ENGINEERING

7 MUNICIPAL OR URBAN ENGINEERING

8 COASTAL STRUCTURES DESIGN AND MANAGEMENT

9 WASTEWATER ENGINEERING

List of Abbreviations

AASHTO	American Association of State Highway and Transportation Officials
AREMA	American Railway Engineering and Maintenance-of-Way Association
ASHE	American Society for Highway Engineering
ASCE	American Society of Civil Engineers
DR	Development Report
CRCP	continuously reinforced concrete pavements
BMET	biomedical equipment technician
BST	Bituminous surface treatment
BLM	Bureau of Land Management
CWA	Clean Water Act
CMMS	computerized maintenance management system
CRCP	continuously reinforced concrete pavements
EDR	electrodialysis reversal
EDM	electronic distance measurement device
EIS	Environmental Impact Statement
EPA	Environmental Protection Agency
GIS	geographical information systems
GLO	General Land Office
GPS	Global Positioning System
HVAC	heating, ventilation and air conditioning
HMA	Hot-Mix Asphalt
IRSE	Institute of Railway Signal Engineers
NZCPS	New Zealand Coastal Policy Statement
ITE	Institute of Transportation Engineers

ICM	Integrated coastal management
ICZM	Integrated Coastal Zone Management
IPCC	Intergovernmental Panel on Climate Change
IAIA	International Association for Impact Assessment
NZCPS	New Zealand Coastal Policy Statement
PCBs	Polychlorinated biphenyl
PCC	Portland cement concrete
PLSS	Public Land Survey System
JPCP	Jointed plain concrete pavement
JRCP	jointed reinforced concrete pavement
MEP	Mechanical, Electrical, and Plumbing
NCEES	National Council of Examiners for Engineering and Surveying
NEPA	National Environmental Policy Act
UNEP	United Nations Environment Programme
TMS	thin membrane surface
WHO	World Health Organization
WWTP	wastewater treatment plant
YPT	Young Professionals in Transportation

Preface

Civil engineering is all about helping people and shaping the world. It is the work that civil engineers do to make our lives much easier. Civil engineering is one of the world's great professions. Civil engineers are responsible for virtually all the great structures around us today and throughout history. The built environment encompasses much of what defines modern civilization. Buildings and bridges are often the first constructions that come to mind, as they are the most conspicuous creations of structural engineering, one of civil engineering's major sub-disciplines. Roads, railroads, subway systems, and airports are designed by transportation engineers, another category of civil engineering. And then there are the less visible creations of civil engineers. Every time you open a water faucet, you expect water to come out, without thinking that civil engineers made it possible. The old civil engineering discipline of sanitary engineering has evolved into modern environmental engineering of such significance that most academic departments have changed their names to civil and environmental engineering. Civil engineers design, build, and maintain the foundation for our modern society - our roads and bridges, drinking water and energy systems, sea ports and airports, and the infrastructure for a cleaner environment, to name just a few. These few examples illustrate that civil engineers do a lot more than design buildings and bridges. They can be found in the aerospace industry, designing jetliners and space stations; in the automotive industry, perfecting the load-carrying capacity of a chassis and improving the crashworthiness of bumpers and doors; and they can be found in the ship building industry, the power industry, and many other industries wherever constructed facilities are involved. And they plan and oversee the construction of these facilities as construction managers.

This text continues to focus on the basics of civil engineering and engineering mechanics to provide students with a balanced and organized approaches.

Content Coverage

Chapter One focuses on architectural engineering concerned with the application of engineering principles to the design and construction of buildings and other structures.

Chapter Two presents the art of determining the relative positions of different object on the surface of the earth by measuring the horizontal distance between them and by preparing a map to any suitable scale.

Chapter Three is structural engineering that focuses on supporting a load safely, and relies on principles of physics and mathematics to design these supports, whereas architectural engineering is concerned with the application of engineering principles to the design and construction of buildings and other structures.

Chapter Four presents the coverage of transportation engineering. Railway, bus, highway, and airport systems are intricate infrastructures that require considerable planning and development.

Chapter Five focuses on practices of road surface engineering, such as paved road surface, should provide a durable, even running surface with adequate skid resistance which is able to withstand the effects of weather. When the road surface condition becomes poor or defective, rehabilitation or resurfacing work is needed to restore the pavement surface to the required level of service.

Chapter Six is concerned with the application of scientific and engineering principles for protection of human populations from the effects of adverse environmental factors; protection of environments, both local and global, from the potentially deleterious effects of natural and human activities; and improvement of environmental quality.

Chapter Seven is concerned with municipal infrastructure that involves specifying, designing, constructing, and maintaining streets, sidewalks, water supply networks, sewers, street lighting, municipal solid waste management and disposal, storage depots for various bulk materials used for maintenance and public works, public parks and cycling infrastructure.

Chapter Eight presents the coastal engineering approaches and issues along coastlines, where constant wave action creates a dynamic setting.

Chapter Nine is focused on the management of water supplies for human and environmental health. It can include everything from sourcing clean supplies of drinking water to the management of wastewater in an environmentally friendly way.

CHAPTER 1

Architectural Engineering

Chapter Contents

- **Introduction**
- **Engineering For Designer**
- **Education**

Architectural engineering is the branch of engineering concerned with the application of engineering principles to the design and construction of buildings and other structures. In architectural engineering, the focus is on the systems used to construct new buildings or restore older ones. These systems include the building structure itself, the lighting, sound, and electrical systems, and the heating, ventilation, and air conditioning systems. Other engineering systems include the plumbing structure, fire safety systems, and energy conservation. Architectural engineers are responsible for coordinating these systems and making sure they are designed and built properly, safely, and economically.

INTRODUCTION

Architectural engineering, also known as building engineering, is the application of engineering principles and technology to building design and construction. Definitions of an architectural engineer may refer to:

- An engineer in the structural, mechanical, electrical, construction or other engineering fields of building design and construction.
- A licensed engineering professional in parts of the United States.

- Architectural engineers are those who work with other engineers and architects for the designing and construction of buildings.

César Pelli's Ratner Athletic Center uses cables and masts as load-bearing devices.

ENGINEERING FOR DESIGNER

Structural Engineering

Structural engineering involves the analysis and design of physical objects (buildings, bridges, equipment supports, towers and walls). Those concentrating on buildings are responsible for the structural performance of a large part of the built environment and are, sometimes, informally referred to as "building engineers". Structural engineers require expertise in strength of materials and in the seismic design of structures covered by earthquake engineering. Architectural Engineers sometimes structural as one aspect of their designs; the structural discipline when practiced as a specialty works closely with architects and other engineering specialists.

Mechanical, Electrical, and Plumbing (MEP)

Mechanical engineering and electrical engineering engineers are specialists, commonly referred to as (mechanical, electrical, and plumbing) when engaged in the building design fields. Also known as "building services engineering" in the United Kingdom, Canada, and Australia. Mechanical engineers often design and oversee the heating, ventilation and air conditioning (HVAC), plumbing, and rain gutter systems. Plumbing designers often include design specifications for simple active fire protection systems, but for more complicated projects, fire protection engineers are often separately retained. Electrical engineers are responsible for the building's power distribution, telecommunication, fire alarm, signalization, lightning protection and control systems, as well as lighting systems.

MEP room in a building

The Architectural Engineer (Pe) in the United States

In many jurisdictions of the United States, the architectural engineer is a licensed engineering professional. Usually a graduate of an architectural engineering university program preparing students to perform whole-building design in competition with architect-engineer teams; or for practice in one of structural, mechanical or electrical fields of building design, but with an appreciation of integrated architectural requirements.

Formal architectural engineering education, following the engineering model of earlier disciplines, developed in the late 19th century, and became widespread in the United States by the mid-20th century. With the establishment of a specific "architectural engineering" NCEES Professional Engineering registration examination in the 1990s, and first offering in April 2003, architectural engineering became recognized as a distinct engineering discipline in the United States. Architectural engineers are not entitled to practice architecture unless they are also licensed as architects.

The Architect as Architectural Engineer

In some countries, the practice of architecture includes planning, designing and overseeing the building's construction, and architecture, as a profession providing architectural services, is referred to as "architectural engineering". In Japan, a "first-class architect" plays the dual role of architect and building engineer, although the services of a licensed "structural design first-class architect" are required for buildings over a certain scale.

In some languages, such as Korean and Arabic, "architect" is literally translated as "architectural engineer". In some countries, an "architectural engineer" (such as the *ingegnere edile* in Italy) is entitled to practice architecture and is often referred to as an architect. These individuals are often also structural engineers. In other countries, such as Germany, Austria, Iran, and most of the Arab countries, architecture graduates receive an engineering degree (*Dipl.-Ing. – Diplom-Ingenieur*).

In Spain, an "architect" has a technical university education and legal powers to carry out building structure and facility projects.

In Brazil, architects and engineers used to share the same accreditation process (CONFEA – Federal Council of Engineering, Architecture and Agronomy). Now the Brazilian architects and urbanists have their own accreditation process (CAU – Architecture and Urbanism Council). Besides traditional architecture design training, Brazilian architecture courses also offer complementary training in engineering disciplines such as structural, electrical, hydraulic and mechanical engineering. After graduation, architects can be responsible for design and construction in these areas (except in electric wiring, where the architect autonomy is limited to systems up to 30kVA), applied to buildings, urban environment, built cultural heritage, landscape planning, interiorscape planning and regional planning.

In Greece licensed architectural engineers are graduates from architecture faculties that belong to the Polytechnic University,obtaining an "Engineering Diploma". They graduate after 5 years of studies and are fully entitled architects once they become members of the Technical Chamber of Greece (TEE – Τεχνικό Επιμελητήριο Ελλάδος). The Engineering Diploma equals a master's degree in ECTS units (300) according to the Bologna Accords.

EDUCATION

The architectural, structural, mechanical and electrical engineering branches each have well established educational requirements that are usually fulfilled by completion of a university program.

An air handling unit is used for the heating and cooling of air in a central location (click on image for legend). Bringing together knowledge of acoustic engineering and HVAC is one example of the multi-disciplined nature of architectural engineering

Architectural Engineering as a Single Integrated Field of Study

What differentiates architectural engineering from architecture (architect) as a separate and single, integrated field of study, compared to other engineering disciplines, is its multi-disciplinary engineering approach. Through training in and appreciation of architecture, the field seeks integration of building systems within its overall building design. Architectural engineering includes the design of building systems including heating, ventilation and air conditioning (HVAC), plumbing, fire protection, electrical, lighting, architectural acoustics, and structural systems. In some university programs, students are required to concentrate on one of the systems; in others, they can receive a generalist architectural or building engineering degree.

REFERENCES

1. Chartered Institute of Building Services Engineers
2. "Licensure". NCEES. Retrieved 2013-10-20.
3. "Architects / Building Engineers in Japan" (PDF). The Japan Architectural Education and Information Center. p. 5. Retrieved 2014-08-16.
4. Edilnews (in Italian)
5. German Chamber of Architects Archived April 27, 2007, at the Wayback Machine.
6. "Ley de Ordenación de la Edificación" (Building Management Act)
7. Resolução 1010/05 – Conselho Federal de Engenharia, Arquitetura e Urbanismo – CONFEA Archived May 15, 2013, at the Wayback Machine.
8. LEI n° 12.378, de 2010 – Presidência da República
9. Polytechnic (Greece)
10. Technical Chamber of Greece (Τεχνικό Επιμελητήριο Ελλάδος)
11. "Role & Objectives". Technical Chamber of Greece. Archived from the original on 2011-08-14.
12. The Technical Chamber of Greece has more than 100,000 members encompassing all the engineering disciplines as well as architecture. A prerequisite for being a member is to be licensed as a qualified engineer or architect and to be a graduate of an engineering and architecture schools of a Greek university, or of an equivalent school from abroad. The Technical Chamber of Greece is the authorized body to provide work licenses to engineers of all disciplines as well as architects, graduated in Greece or abroad. The license is awarded after examinations. The examinations take place three to four times a year.
13. Bologna Process#Qualifications Framework of the European Higher Education Area

CHAPTER 2

Surveying

Chapter Contents

- **Introduction**
- **Definitions**
- **Surveying Techniques**

It is the art of determining the relative positions of different object on the surface of the earth by measuring the horizontal distance between them and by preparing a map to any suitable scale. Thus, in this process, the measurements are taken only in the horizontal plane.

INTRODUCTION

Surveying or land surveying is the technique, profession, and science of determining the terrestrial or three-dimensional position of points and the distances and angles between them. A land surveying professional is called a land surveyor. These points are usually on the surface of the Earth, and they are often used to establish land maps and boundaries for ownership, locations like building corners or the surface location of subsurface features, or other purposes required by government or civil law, such as property sales.

Surveyors work with elements of geometry, trigonometry, regression analysis, physics, engineering, metrology, programming languages and the law. They use equipment like total stations, robotic total stations, GPS receivers, retroreflectors, 3D scanners, radios, handheld tablets, digital levels, drones, GIS and surveying software.

Surveying has been an element in the development of the human environment since the beginning of recorded history. The planning and execution of most forms of construction require it. It is also used in transport, communications, mapping, and the definition of legal boundaries for land ownership. It is an important tool for research in many other scientific disciplines.

A surveyor at work with an infrared reflector used for distance measurement.

DEFINITIONS

ACSM

The American Congress on Surveying and Mapping (ACSM), defines surveying as the science and art of making *all* essential measurements to determine the relative position of points or physical and cultural details above, on, or beneath the surface of the Earth, and to depict them in a usable form, or to establish the position of points or details.

Also per ACSM, the type of surveying known as "land surveying" is the detailed study or inspection, as by gathering information through observations, measurements in the field, questionnaires, or research of legal instruments, and data analysis in the support of planning, designing, and establishing of property boundaries. It involves the re-establishment of cadastral surveys and land boundaries based on documents of record and historical evidence, as well as certifying surveys (as required by statute or local ordinance) of subdivision plats or maps, registered land surveys, judicial surveys, and space delineation. Land surveying can include associated services such as mapping and related data accumulation, construction layout surveys, precision measurements of length, angle, elevation, area, and volume, as well as horizontal and vertical control surveys, and the analysis and utilization of land survey data.

FIG

The International Federation of Surveyors defines the function of surveying as:

A surveyor is a professional person with the academic qualifications and technical expertise to conduct one, or more, of the following activities;

- to determine, measure and represent land, three-dimensional objects, point-fields and trajectories;
- to assemble and interpret land and geographically related information,
- to use that information for the planning and efficient administration of the land, the sea and any structures thereon; and,
- to conduct research into the above practices and to develop them.

Ancient Surveying

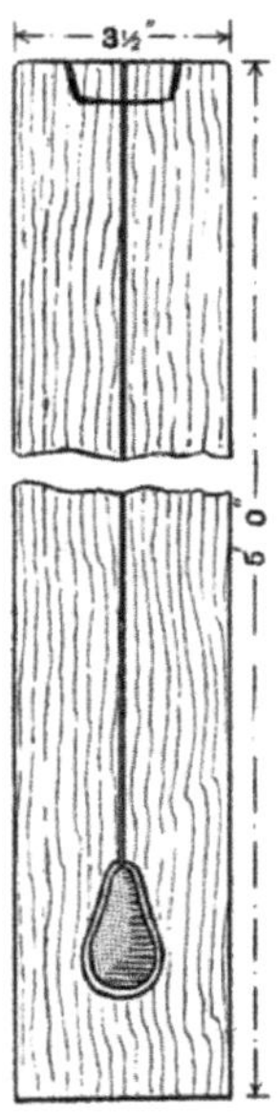

A plumb rule from the book Cassells' Carpentry and Joinery

Basic surveyance has occurred since humans built the first large structures. The prehistoric monument at Stonehenge (c. 2500 BC) was set out by prehistoric surveyors using peg and rope geometry.

In ancient Egypt, a rope stretcher would use simple geometry to re-establish boundaries after the annual floods of the Nile River. The almost perfect squareness and north-south orientation of the Great Pyramid of Giza, built c. 2700 BC, affirm the Egyptians' command of surveying. The *Groma* instrument originated in Mesopotamia (early 1st millennium BC).

The mathematician Liu Hui described ways of measuring distant objects in his work *Haidao suanjing* or *The Sea Island Mathematical Manual*, published in 263 AD.

The Romans recognized land surveyors as a profession. They established the basic measurements under which the Roman Empire was divided, such as a tax register of conquered lands (300 AD). Roman surveyors were known as *Gromatici*.

In medieval Europe, beating the bounds maintained the boundaries of a village or parish. This was the practice of gathering a group of residents and walking around the parish or village to establish a communal memory of the boundaries. Young boys were included to ensure the memory lasted as long as possible.

In England, William the Conqueror commissioned the Domesday Book in 1086. It recorded the names of all the land owners, the area of land they owned, the quality of the land, and specific information of the area's content and inhabitants. It did not include maps showing exact locations.

Modern Surveying

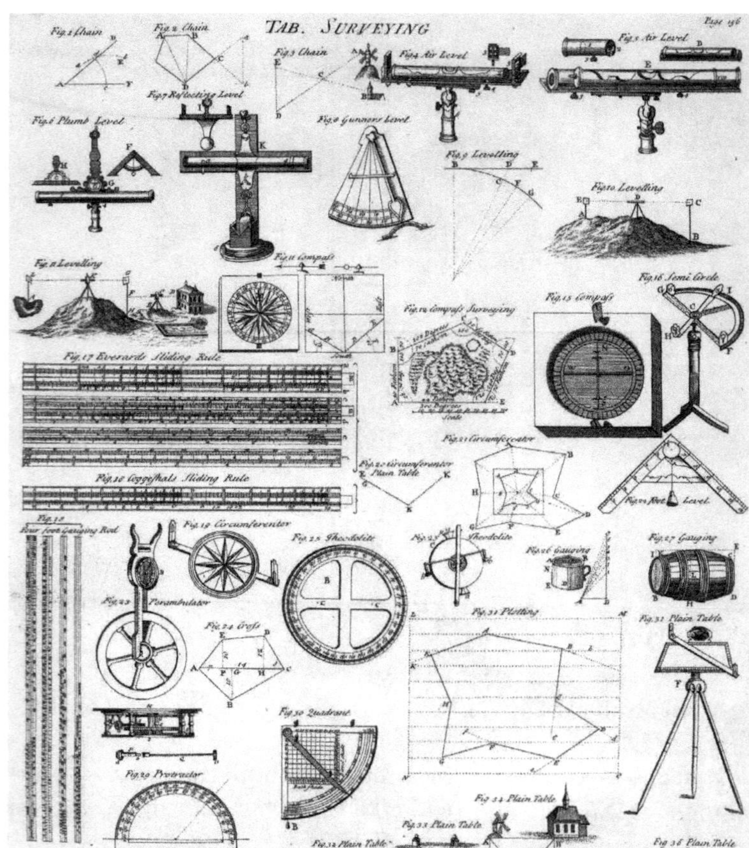

Table of Surveying, 1728 *Cyclopaedia*

Abel Foullon described a plane table in 1551, but it is thought that the instrument was in use earlier as his description is of a developed instrument.

Gunter's chain was introduced in 1620 by English mathematician Edmund Gunter. It enabled plots of land to be accurately surveyed and plotted for legal and commercial purposes.

Leonard Digges described a Theodolite that measured horizontal angles in his book *A geometric practice named Pantometria* (1571). Joshua Habermel (de:Erasmus Habermehl) created a theodolite with a compass and tripod in 1576. Johnathon Sission was the first to incorporate a telescope on a theodolite in 1725.

In the 18th century, modern techniques and instruments for surveying began to be used. Jesse Ramsden introduced the first precision theodolite in 1787. It was an instrument for measuring angles in the horizontal and vertical planes. He created his great theodolite using an accurate dividing engine of his own design. Ramsden›s theodolite represented a great step forward in the instrument›s accuracy. William Gascoigne invented an instrument that used a telescope with an installed crosshair as a target device, in 1640. James Watt developed an optical meter for the measuring of distance in 1771; it measured the parallactic angle from which the distance to a point could be deduced.

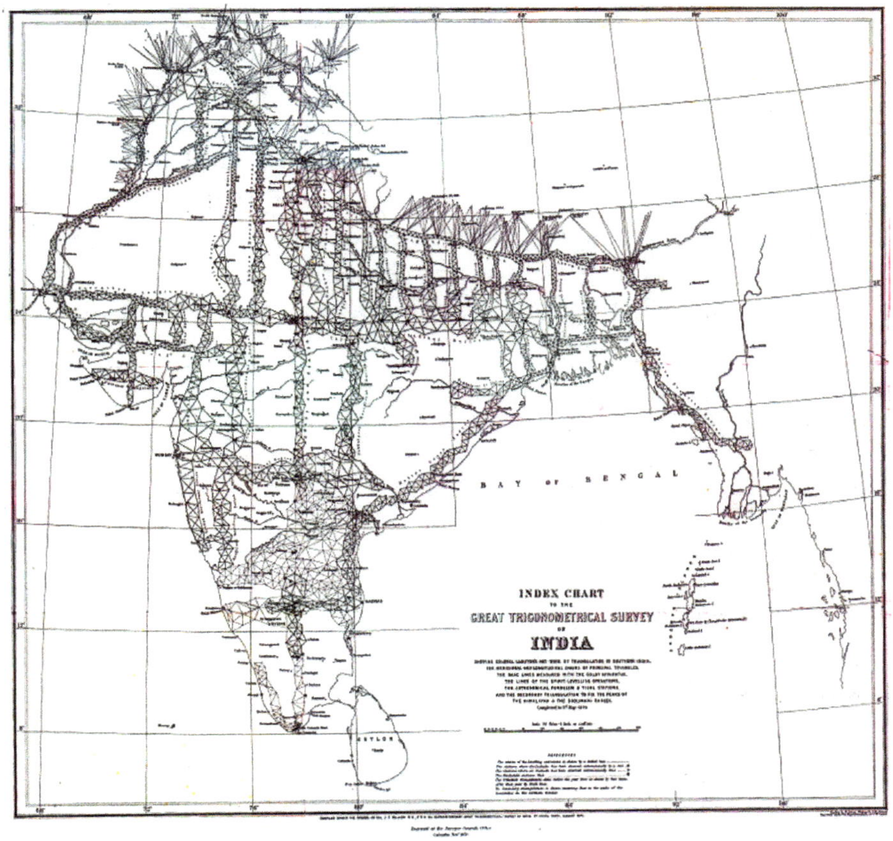

A map of India showing the Great Trigonometrical Survey, produced in 1870

Dutch mathematician Willebrord Snellius (a.k.a. Snel van Royen) introduced the modern systematic use of triangulation. In 1615 he surveyed the distance from Alkmaar to Breda, approximately 72 miles (116,1 kilometres). He underestimated this distance by 3.5%. The survey was a chain of quadrangles containing 33 triangles in all. Snell showed how planar formulae could be corrected to allow for the curvature of the earth. He also showed how to resection, or calculate, the position of a point inside a triangle using the angles cast between the vertices at the unknown point. These could be measured more accurately than bearings of the vertices, which depended on a compass. His work established the idea of surveying a primary network of control points, and locating subsidiary points inside the primary network later. Between 1733 and 1740, Jacques Cassini and his son César undertook the first triangulation of France. They included a re-surveying of the meridian arc, leading to the publication in 1745 of the first map of France constructed on rigorous principles. By this time, triangulation methods were by then well established for local map-making.

It was only towards the end of the 18th century that detailed triangulation network surveys mapped whole countries. In 1784, a team from General William Roy's Ordnance Survey of Great Britain began the Principal Triangulation of Britain. The first Ramsden theodolite was built for this survey. The survey was finally completed in 1853. The Great Trigonometric Survey of India began in 1801. The Indian survey had an enormous scientific impact. It was responsible for one of the first accurate measurements of a section of an arc of longitude, and for measurements of the geodesic anomaly. It named and mapped Mount Everest and the other Himalayan peaks. Surveying became a professional occupation in high demand at the turn of the 19th century with the onset of the Industrial Revolution. The profession developed more accurate instruments to aid its work. Industrial infrastructure projects used surveyors to lay out canals, roads and rail.

In the US, the Land Ordinance of 1785 created the Public Land Survey System. It formed the basis for dividing the western territories into sections to allow the sale of land. The PLSS divided states into township grids which were further divided into sections and fractions of sections.

Napoleon Bonaparte founded continental Europe's first cadastre in 1808. This gathered data on the number of parcels of land, their value, land usage, and names. This system soon spread around Europe.

A railroad surveying party at Russel›s Tank, Arizona in the 1860s

Robert Torrens introduced the Torrens system in South Australia in 1858. Torrens intended to simplify land transactions and provide reliable titles via a centralized register of land. The Torrens system was adopted in several other nations of the English-speaking world. Surveying became increasingly important with the arrival of railroads in the 1800s. Surveying was necessary so that railroads could plan technologically and financially viable routes.

20th Century

A German engineer surveying during the First World War, 1918

At the beginning of the century surveyors had improved the older chains and ropes, but still faced the problem of accurate measurement of long distances. Dr Trevor Lloyd Wadley developed the Tellurometer during the 1950s. It measures long distances using two microwave transmitter/receivers. During the late 1950s Geodimeter introduced electronic distance measurement (EDM) equipment. EDM units use a multi frequency phase shift of light waves to find a distance. These instruments saved the need for days or weeks of chain measurement by measuring between points kilometers apart in one go.

Advances in electronics allowed miniaturization of EDM. In the 1970s the first instruments combining angle and distance measurement appeared, becoming known as total stations. Manufacturers added more equipment by degrees, bringing improvements in accuracy and speed of measurement. Major advances include tilt compensators, data recorders, and on-board calculation programs.

The first Satellite positioning system was the US Navy TRANSIT system. The first successful launch took place in 1960. The system's main purpose was to provide position information to Polaris missile submarines. Surveyors found they could use field receivers to determine the location of a point. Sparse satellite cover and large equipment made observations laborious, and inaccurate. The main use was establishing benchmarks in remote locations.

The US Air Force launched the first prototype satellites of the Global Positioning System (GPS) in 1978. GPS used a larger constellation of satellites and improved signal transmission to provide more accuracy. Early GPS observations required several hours of observations by a static receiver to reach survey accuracy requirements. Recent improvements to both satellites and receivers allow Real Time Kinematic (RTK) surveying. RTK surveys get high-accuracy measurements by using a fixed base station and a second roving antenna. The position of the roving antenna can be tracked.

21st Century

The theodolite, total station, and RTK GPS survey remain the primary methods in use.

Remote sensing and satellite imagery continue to improve and become cheaper, allowing more commonplace use. Prominent new technologies include three-dimensional (3D) scanning and use of lidar for topographical surveys. UAV technology along with photogrammetric image processing is also appearing.

Surveying Equipment

Hardware

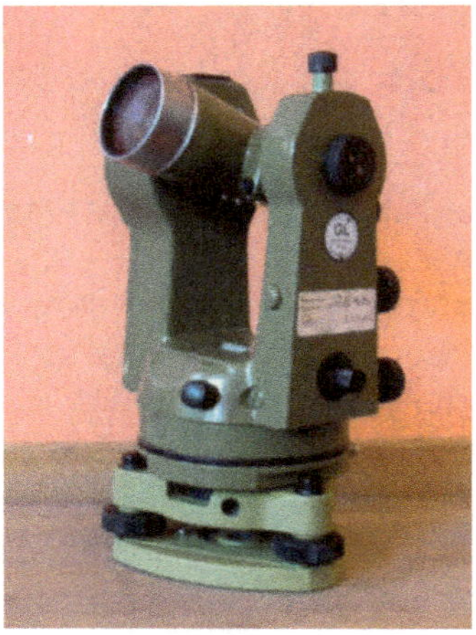

Surveying equipment. Clockwise from upper left: optical theodolite, robotic total station, RTK GPS base station, optical level.

The main surveying instruments in use around the world are the theodolite, measuring tape, total station, 3D scanners, GPS/GNSS, level and rod. Most instruments screw onto a tripod when in use. Tape measures are often used for measurement of smaller distances. 3D scanners and various forms of aerial imagery are also used.

The theodolite is an instrument for the measurement of angles. It uses two separate *circles*, *protractors* or *alidades* to measure angles in the horizontal and the vertical plane.

A telescope mounted on trunnions is aligned vertically with the target object. The whole upper section rotates for horizontal alignment. The vertical circle measures the angle that the telescope makes against the vertical, known as the zenith angle. The horizontal circle uses an upper and lower plate. When beginning the survey, the surveyor points the instrument in a known direction (bearing), and clamps the lower plate in place. The instrument can then rotate to measure the bearing to other objects. If no bearing is known or direct angle measurement is wanted, the instrument can be set to zero during the initial sight. It will then read the angle between the initial object, the theodolite itself, and the item that the telescope aligns with.

The gyrotheodolite is a form of theodolite that uses a gyroscope to orient itself in the absence of reference marks. It is used in underground applications.

The total station is a development of the theodolite with an electronic distance measurement device (EDM). A total station can be used for leveling when set to the horizontal plane. Since their introduction, total stations have shifted from optical-mechanical to fully electronic devices.

Modern top-of-the-line total stations no longer need a reflector or prism to return the light pulses used for distance measurements. They are fully robotic, and can even e-mail point data to a remote computer and connect to satellite positioning systems, such as Global Positioning System. Real time kinematic GPS systems have increased the speed of surveying, but they are still only horizontally accurate to about 20 mm and vertically to 30–40 mm.

GPS surveying differs from other GPS uses in the equipment and methods used. Static GPS uses two receivers placed in position for a considerable length of time. The long span of time lets the receiver compare measurements as the satellites orbit. The changes as the satellites orbit also provide the measurement network with well conditioned geometry. This produces an accurate baseline that can be over 20 km long. RTK surveying uses one static antenna and one roving antenna. The static antenna tracks changes in the satellite positions and atmospheric conditions. The surveyor uses the roving antenna to measure the points needed for the survey. The two antennas use a radio link that allows the static antenna to send corrections to the roving antenna. The roving antenna then applies those corrections to the GPS signals it is receiving to calculate its own position. RTK surveying covers smaller distances than static methods. This is because divergent conditions further away from the base reduce accuracy.

Surveying instruments have characteristics that make them suitable for certain uses. Theodolites and levels are often used by constructors rather than surveyors in first world countries. The constructor can perform simple survey tasks using a relatively cheap instrument. Total stations are workhorses for many professional surveyors because they are versatile and reliable in all conditions. The productivity improvements from a GPS on large scale surveys makes them popular for major infrastructure or data gathering projects. One-person robotic-guided total stations allow surveyors to measure without extra workers to aim the telescope or record data. A fast but expensive way to measure large areas is with a helicopter, using a GPS to record the location of the helicopter and a laser scanner to measure the ground. To increase precision, surveyors place beacons on the ground (about 20 km (12 mi) apart). This method reaches precisions between 5–40 cm (depending on flight height).

Surveyors use ancillary equipment such as tripods and instrument stands; staves and beacons used for sighting purposes; PPE; vegetation clearing equipment; digging implements for finding survey markers buried over time; hammers for placements of markers in various surfaces and structures; and portable radios for communication over long lines of sight.

Software

Land surveyors, construction professionals and civil engineers using total station, GPS, 3D scanners and other collector data use Land Surveying Software to increase efficiency, accuracy and productivity. Land Surveying Software is a staple of contemporary land surveying.

SURVEYING TECHNIQUES

A standard Brunton Geo compass, still used commonly today by geographers, geologists and surveyors for field-based measurements

Surveyors determine the position of objects by measuring angles and distances. The factors that can affect the accuracy of their observations are also measured. They then use this data to create vectors, bearings, co-ordinates, elevations, areas, volumes, plans and maps. Measurements are often split into horizontal and vertical components to simplify calculation. GPS and astronomic measurements also need measurement of a time component.

Distance Measurement

Example of modern equipment for surveying (Field-Map technology): GPS, laser rangefinder and field computer allows surveying as well as cartography (creation of map in real-time) and field data collection.

Before EDM devices, distances were measured using a variety of means. These included chains with links of a known length such as a Gunter's chain, or measuring tapes made of steel or invar. To measure horizontal distances, these chains or tapes were pulled taut to reduce sagging and slack. The distance had to be adjusted for heat expansion. Attempts to hold the measuring instrument level would also be made. When measuring up a slope, the surveyor might have to "break" (break chain) the measurement- use an increment less than the total length of the chain. Perambulators, or measuring wheels, were used to measure longer distances but not to a high level of accuracy. Tacheometry is the science of measuring distances by measuring the angle between two ends of an object with a known size. It was sometimes used before to the invention of EDM where rough ground made chain measurement impractical.

Angle Measurement

Historically, horizontal angles were measured by using a compass to provide a magnetic bearing or azimuth. Later, more precise scribed discs improved angular resolution. Mounting telescopes with reticles atop the disc allowed more precise sighting (see theodolite). Levels and calibrated circles allowed measurement of vertical angles. Verniers allowed measurement to a fraction of a degree, such as with a turn-of-the-century transit.

The plane table provided a graphical method of recording and measuring angles, which reduced the amount of mathematics required. In 1829 Francis Ronalds invented a reflecting instrument for recording angles graphically by modifying the octant.

By observing the bearing from every vertex in a figure, a surveyor can measure around the figure. The final observation will be between the two points first observed, except with a 180° difference. This is called a *close*. If the first and last bearings are different, this shows the error in the survey, called the *angular misclose*. The surveyor can use this information to prove that the work meets the expected standards.

Levelling

Center for Operational Oceanographic Products and Services staff member conducts tide station leveling in support of the US Army Corp of Engineers in Richmond, Maine.

The simplest method for measuring height is with an altimeter using air pressure to find height. When more precise measurements are needed, means like precise levels (also known as differential leveling) are used. When precise leveling, a series of measurements between two points are taken using an instrument and a measuring rod. Differences in height between the measurements are added and subtracted in a series to get the net difference in elevation between the two endpoints. With the Global Positioning System (GPS), elevation can be measured with satellite receivers. Usually GPS is somewhat less accurate than traditional precise leveling, but may be similar over long distances.

When using an optical level, the endpoint may be out of the effective range of the instrument. There may be obstructions or large changes of elevation between the endpoints. In these situations, extra setups are needed. *Turning* is a term used when referring to moving the level to take an elevation shot from a different location. To "turn" the level, one must first take a reading and record the elevation of the point the rod is located on. While the rod is being kept in exactly the same location, the level is moved to a new location where the rod is still visible. A reading is taken from the new location of the level and the height difference is used to find the new elevation of the level gun. This is repeated until the series of measurements is completed. The level must be horizontal to get a valid measurement. Because of this, if the horizontal crosshair of the instrument is lower than the base of the rod, the surveyor will not be able to sight the rod and get a reading. The rod can usually be raised up to 25 feet high, allowing the level to be set much higher than the base of the rod.

Determining Position

The primary way of determining one's position on the earth's surface when no known positions are nearby is by astronomic observations. Observations to the sun, moon and stars could all be made using navigational techniques. Once the instrument's position and bearing to a star is determined, the bearing can be transferred to a reference point on the earth. The point can then be used as a base for further observations. Survey-accurate astronomic positions were difficult to observe and calculate and so tended to be a base off which many other measurements were made. Since the advent of the GPS system, astronomic observations are rare as GPS allows adequate positions to be determined over most of the surface of the earth.

Reference Networks

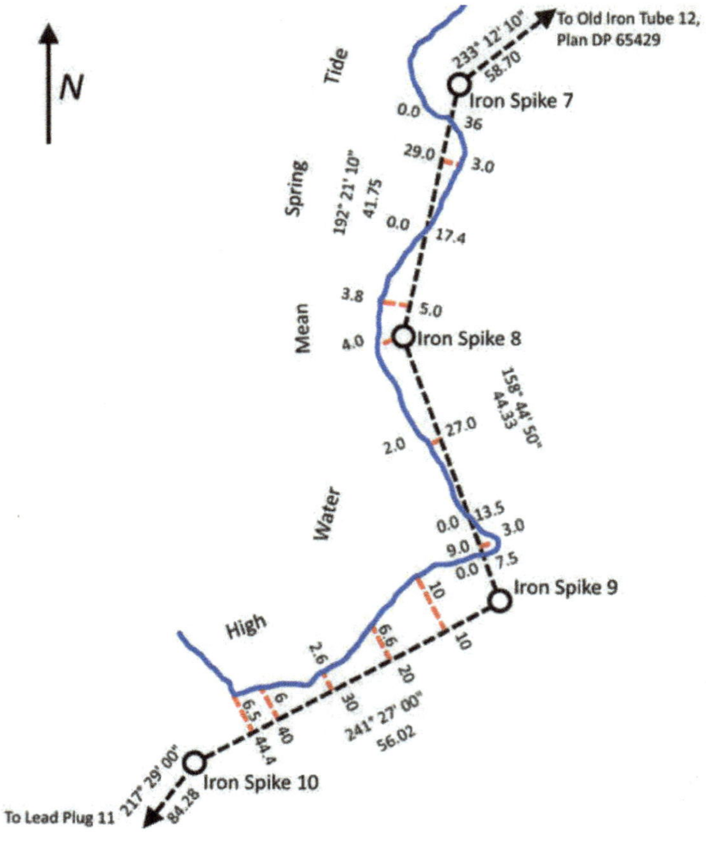

A survey using traverse and offset measurements to record the location of the shoreline shown in blue. Black dashed lines are traverse measurements between reference points (black circles). The red lines are offsets measured at right angles to the traverse lines.

Few survey positions are derived from first principles. Instead, most surveys points are measured relative to previous measured points. This forms a reference or *control* network where each point can be used by a surveyor to determine their own position when beginning a new survey.

Survey points are usually marked on the earth›s surface by objects ranging from small nails driven into the ground to large beacons that can be seen from long distances. The surveyors can set up their instruments on this position and measure to nearby objects. Sometimes a tall, distinctive feature such as a steeple or radio aerial has its position calculated as a reference point that angles can be measured against.

Triangulation is a method of horizontal location favoured in the days before EDM and GPS measurement. It can determine distances, elevations and directions between distant objects. Since the early days of surveying, this was the primary method of determining accurate positions of objects for topographic maps of large areas. A surveyor first needs to know the horizontal distance between two of the objects, known as the baseline. Then the heights, distances and angular position of other objects can be derived, as long as they are visible from one of the original objects. High-accuracy transits or theodolites were used, and angle measurements repeated for increased accuracy. See also Triangulation in three dimensions.

Offsetting is an alternate method of determining position of objects, and was often used to measure imprecise features such as riverbanks. The surveyor would mark and measure two known positions on the ground roughly parallel to the feature, and mark out a baseline between them. At regular intervals, a distance was measured at right angles from the first line to the feature. The measurements could then be plotted on a plan or map, and the points at the ends of the offset lines could be joined to show the feature.

Traversing is a common method of surveying smaller areas. The surveyor starts from an old reference mark or known position and places a network of reference marks covering the survey area. They then measure bearings and distances between the reference marks, and to the target features. Most traverses form a loop pattern or link between two prior reference marks so the surveyor can check their measurements.

Datum and Coordinate Systems

Many surveys do not calculate positions on the surface of the earth, but instead measure the relative positions of objects. However, often the surveyed items need to be compared to outside data, such as boundary lines or previous surveys objects. The oldest way of describing a position is via latitude and longitude, and often a height above sea level. As the surveying profession grew it created Cartesian coordinate systems to simplify the mathematics for surveys over small parts of the earth. The simplest coordinate systems assume that the earth is flat and measure from an arbitrary point, known as a 'datum' (singular form of data). The coordinate system allows easy calculation of the distances and direction between objects over small areas. Large areas distort due to the earth's curvature. North is often defined as true north at the datum.

For larger regions, it is necessary to model the shape of the earth using an ellipsoid or a geoid. Many countries have created coordinate-grids customized to lessen error in their area of the earth.

Errors and Accuracy

A basic tenet of surveying is that no measurement is perfect, and that there will always be a small amount of error. There are three classes of survey errors:

- *Gross errors or blunders:* Errors made by the surveyor during the survey. Upsetting the instrument, misaiming a target, or writing down a wrong measurement are all gross errors. A large gross error may reduce the accuracy to an unacceptable level. Therefore, surveyors use redundant measurements and independent checks to detect these errors early in the survey.
- *Systematic:* Errors that follow a consistent pattern. Examples include effects of temperature on a chain or EDM measurement, or a poorly adjusted spirit-level causing a tilted instrument or target pole. Systematic errors that have known effects can be compensated or corrected.
- *Random:* Random errors are small unavoidable fluctuations. They are caused by imperfections in measuring equipment, eyesight, and conditions. They can be minimized by redundancy of measurement and avoiding unstable conditions. Random errors tend to cancel each other out, but checks must be made to ensure they are not propagating from one measurement to the next.

Surveyors avoid these errors by calibrating their equipment, using consistent methods, and by good design of their reference network. Repeated measurements can be averaged and any outlier measurements discarded. Independent checks like measuring a point from two or more locations or using two different methods are used. Errors can be detected by comparing the results of the two measurements.

Once the surveyor has calculated the level of the errors in his work, it is adjusted. This is the process of distributing the error between all measurements. Each observation is weighted according to how much of the total error it is likely to have caused and part of that error is allocated to it in a proportional way. The most common methods of adjustment are the Bowditch method, also known as the compass rule, and the Principle of least squares method.

The surveyor must be able to distinguish between accuracy and precision. In the United States, surveyors and civil engineers use units of feet wherein a survey foot breaks down into 10ths and 100ths. Many deed descriptions containing distances are often expressed using these units (125.25 ft). On the subject of accuracy, surveyors are often held to a standard of one one-hundredth of a foot; about 1/8 inch. Calculation and mapping tolerances are much smaller wherein achieving near-perfect closures are desired. Though tolerances will vary from project to project, in the field and day to day usage beyond a 100th of a foot is often impractical.

Types of Surveys

Local organisations or regulatory bodies class specializations of surveying in different ways. Broad groups are:

- *As-built survey*: a survey that documents the location of recently constructed elements of a construction project. Asbuilt surveys are done for record, completion evaluation and payment purposes. An as-built survey is also known as a 'works as executed survey'. As built surveys are often presented in red or redline and laid over existing plans for comparison with design information.

- *Cadastral or boundary surveying*: a survey that establishes or re-establishes boundaries of a parcel using a legal description. It involves the setting or restoration of monuments or markers at the corners or along the lines of the parcel. These take the form of iron rods, pipes, or concrete monuments in the ground, or nails set in concrete or asphalt. The *ALTA/ACSM* Land Title Survey is a standard proposed by the American Land Title Association and the American Congress on Surveying and Mapping. It incorporates elements of the boundary survey, mortgage survey, and topographic survey.
- *Control surveying*: Control surveys establish reference points to use as starting positions for future surveys. Most other forms of surveying will contain elements of control surveying.
- *Construction surveying*
- *Deformation survey:* a survey to determine if a structure or object is changing shape or moving. First the positions of points on an object are found. A period of time is allowed to pass and the positions are then re-measured and calculated. Then a comparison between the two sets of positions is made.
- *Dimensional control survey:* This is a type of survey conducted in or on an non-level surface. Common in the oil and gas industry to replace old or damaged pipes on a like-for-like basis. The advantage of dimensional control survey is that the instrument used to conduct the survey does not need to be level. This is useful in the off-shore industry, as not all platforms are fixed and are thus subject to movement.
- *Engineering surveying*: topographic, layout, and as-built surveys associated with engineering design. They often need geodetic computations beyond normal civil engineering practice.
- *Foundation survey*: a survey done to collect the positional data on a foundation that has been poured and is cured. This is done to ensure that the foundation was constructed in the location, and at the elevation, authorized in the *plot plan*, *site plan*, or *subdivision plan*.
- *Hydrographic survey*: a survey conducted with the purpose of mapping the shoreline and bed of a body of water. Used for navigation, engineering, or resource management purposes.
- *Leveling*: either finds the elevation of a given point or establish a point at a given elevation.
- *LOMA survey*: Survey to change base flood line, removing property from a SFHA special flood hazard area.
- *Measured survey* : a building survey to produce plans of the building. such a survey may be conducted before renovation works, for commercial purpose, or at end of the construction process.

- *Mining surveying*: Mining surveying includes directing the digging of mine shafts and galleries and the calculation of volume of rock. It uses specialised techniques due to the restraints to survey geometry such as vertical shafts and narrow passages.
- *Mortgage survey:* A *mortgage survey* or *physical survey* is a simple survey that delineates land boundaries and building locations. It checks for *encroachment*, building setback restrictions and shows nearby flood zones. In many places a mortgage survey is a precondition for a mortgage loan.
- *Photographic control survey*: A survey that creates reference marks visible from the air to allow aerial photographs to be rectified.
- *Stakeout, Layout or Setout*: an element of many other surveys where the calculated or proposed position of an object is marked on the ground. This can be temporary or permanent. This is an important component of engineering and cadastral surveying.
- *Structural survey*: a detailed inspection to report upon the physical condition and structural stability of a building or structure. It highlights any work needed to maintain it in good repair.
- *Subdivision*: A boundary survey that splits a property into two or more smaller properties.
- *Topographic survey*: a survey that measures the elevation of points on a particular piece of land, and presents them as contour lines on a plot.

Plane and Geodetic Surveying

Based on the considerations and true shape of the earth, surveying is broadly classified into two types.

Plane surveying assumes the earth is flat. Curvature and spheroidal shape of the earth is neglected. In this type of surveying all triangles formed by joining survey lines are considered as plane triangles. It is employed for small survey works where errors due to the earth's shape are too small to matter.

In *geodetic surveying* the curvature of the earth is taken into account while calculating reduced levels, angles, bearings and distances. This type of surveying is usually employed for large survey works. Survey works up to 100 square miles (260 square kilometers) are treated as plane and beyond that are treated as geodetic. In geodetic surveying necessary corrections are applied to reduced levels, bearings and other observations.

The Surveying Profession

The pundit (explorer) cartographer Nain Singh Rawat (19th century) received a Royal Geographical Society gold medal in 1876, for his efforts in exploring the Himalayas for the British

An all-female surveying crew in Idaho, 1918

The basic principles of surveying have changed little over the ages, but the tools used by surveyors have evolved. Engineering, especially civil engineering, often needs surveyors.

Surveyors help determine the placement of roads, railways, reservoirs, dams, pipelines, retaining walls, bridges, and buildings. They establish the boundaries of legal descriptions and political divisions. They also provide advice and data for *geographical information systems* (GIS) that record land features and boundaries.

Surveyors must have a thorough knowledge of algebra, basic calculus, geometry, and trigonometry. They must also know the laws that deal with surveys, real property, and contracts.

Most jurisdictions recognize three different levels of qualification:

Survey assistants or *chainmen* are usually unskilled workers who help the surveyor. They place target reflectors, find old reference marks, and mark points on the ground. The term 'chainman' derives from past use of measuring chains. An assistant would move the far end of the chain under the surveyor's direction.

Survey technicians often operate survey instruments, run surveys in the field, do survey calculations, or draft plans. A technician usually has no legal authority and cannot certify his work. Not all technicians are qualified, but qualifications at the certificate or diploma level are available.

Licensed, registered, or chartered surveyors usually hold a degree or higher qualification. They are often required to pass further exams to join a professional association or to gain certifying status. Surveyors are responsible for planning and management of surveys. They have to ensure that their surveys, or surveys performed under their supervision, meet the legal standards. Many principals of surveying firms hold this status.

Informal Surveying

Not all surveys are carried out by professional surveyors. Depending on the jurisdiction and circumstances, the builders of a structure may set it out themselves. Surveyors often set out the most significant corners of a building. The builders then lay out the rest of the building themselves using simple surveying techniques.

Licensing

Licensing requirements vary with jurisdiction, and are commonly consistent within national borders. Prospective surveyors usually have to receive a degree in surveying, followed by a detailed examination of their knowledge of surveying law and principles specific to the region they wish to practice in, and undergo a period of on-the-job training or portfolio building before they are awarded a license to practise. Licensed surveyors usually receive a post nominal, which varies depending on where they qualified. The system has replaced older apprenticeship systems.

A licensed land surveyor is generally required to sign and seal all plans. The state dictates the format, showing their name and registration number.

In many jurisdictions, surveyors must mark their registration number on survey monuments when setting boundary corners. Monuments take the form of capped iron rods, concrete monuments, or nails with washers.

Surveying Institutions

Surveying students with their professor at the Helsinki University of Technology in the late 19th century

Most countries' governments regulate at least some forms of surveying. Their survey agencies establish regulations and standards. Standards control accuracy, surveying credentials, monumentation of boundaries and maintenance of geodetic networks. Many nations devolve this authority to regional entities or states/provinces. Cadastral surveys tend to be the most regulated because of the permanence of the work. Lot boundaries established by cadastral surveys may stand for hundreds of years without modification.

Most jurisdictions also have a form of professional institution representing local surveyors. These institutes often endorse or license potential surveyors, as well as set and enforce ethical standards. The largest institution is the International Federation of Surveyors (Abbreviated FIG, for French: *Fédération Internationale des Géomètres*). They represent the survey industry worldwide.

Building Surveying

Most English-speaking countries consider building surveying a distinct profession. They have their own professional associations and licensing requirements. Building surveyors focus on investigating the condition of buildings as well as legal compliance work.

Cadastral Surveying

One of the primary roles of the land surveyor is to determine the boundary of real property on the ground. The surveyor must determine where the adjoining landowners wish to put the boundary. The boundary is established in legal documents and plans prepared by attorneys, engineers, and land surveyors. The surveyor then puts monuments on the corners of the new boundary. They might also find or resurvey the corners of the property monumented by prior surveys.

Cadastral land surveyors are licensed by governments. The cadastral survey branch of the Bureau of Land Management (BLM) conducts most cadastral surveys in the United States. They consult with Forest Service, National Park Service, Army Corps of Engineers, Bureau of Indian Affairs, Fish and Wildlife Service, Bureau of Reclamation, and others. The BLM used to be known as the General Land Office (GLO).

In states organized per the Public Land Survey System (PLSS), surveyors must carry out BLM cadastral surveys under that system.

Cadastral surveyors often have to work around changes to the earth that obliterate or damage boundary monuments. When this happens, they must consider evidence that is not recorded on the title deed. This is known as extrinsic evidence.

REFERENCES

1. "Definition". fig.net. Retrieved 2016-02-17.
2. Johnson, Anthony, Solving Stonehenge: The New Key to an Ancient Enigma. (Thames & Hudson, 2008) ISBN 978-0-500-05155-9
3. Hong-Sen Yan & Marco Ceccarelli (2009), International Symposium on History of Machines and Mechanisms: Proceedings of HMM 2008, Springer, p. 107, ISBN 1-4020-9484-1
4. Lewis, M. J. T. (2001-04-23). Surveying Instruments of Greece and Rome. Cambridge University Press. ISBN 9780521792974. Retrieved 30 August 2012.
5. Turner, Gerard L'E. Nineteenth Century Scientific Instruments, Sotheby Publications, 1983, ISBN 0-85667-170-3
6. Sturman, Brian; Wright, Alan. "The History of the Tellurometer" (PDF). International Federation of Surveyors. Retrieved 20 July 2014.
7. Cheves, Marc. "Geodimeter-The First Name in EDM". Archived from the original on March 10, 2014. Retrieved 2014-07-20.
8. Mahun, Jerry. "Electronic Distance Measurement". Jerrymahun.com. Retrieved 2014-07-20.
9. National Cooperative Highway Research Program: Collecting, Processing and Integrating GPS data into GIS, p. 40. Published by Transportation Research Board, 2002 ISBN 0-309-06916-5 ISBN 978-0-309-06916-8
10. Toni Schenk, Suyoung Seo, Beata Csatho:Accuracy Study of Airborne Laser Scanning Data with Photogrammetry, p. 118 ArchivedMarch 25, 2009, at the Wayback Machine.
11. "View DigitalGlobe Imagery Solutions @ Geospatial Forum".
12. Ronalds, B.F. (2016). Sir Francis Ronalds: Father of the Electric Telegraph. London: Imperial College Press. ISBN 978-1-78326-917-4.
13. Kahmen, Heribert; Faig, Wolfgang (1988). Surveying. Berlin: de Gruyter. p. 9. ISBN 3-11-008303-5. Retrieved 2014-08-10.
14. BC Punmia. Surveying by BC Punmia. p. 2. Retrieved 9 December 2014h.

15. N N Basak. Surveying and Levelling. p. 542. Retrieved 28 July 2016.
16. BC Punmia. Surveying by BC Punmia. p. 2. Retrieved 9 December 2014.
17. A History of the Rectangular Survey System by C. Albert White, 1983, Pub: Washington, D.C. : U.S. Dept. of the Interior, Bureau of Land Management : For sale by Supt. of Docs., U.S. G.P.O.,
18. Richards, D., & Hermansen, K. (1995). Use of extrinsic evidence to aid interpretation of deeds. Journal of Surveying Engineering, (121), 178.

CHAPTER 3

Structural Engineering: Elements and Systems

Chapter Contents

- **Introduction**
- **Structural Engineer (Professional)**
- **Specializations**
- **Structural Element**
- **Beam (Structure)**
- **Strut**
- **Tie (Engineering)**
- **Thin-Shell Structure**
- **Concrete Slab**
- **Deck (Bridge)**
- **Structural System**

Structural engineering is the design of structural support systems for buildings, bridges, earthworks, and industrial structures. This branch of engineering focuses on supporting a load safely, and relies on principles of physics and mathematics to design these supports. The work is carried out by structural engineers, though in many areas, the lines between civil and structural engineering are blurred. As buildings become increasingly complicated, many universities and licensing boards are creating separate programs and certifications for structural engineers, helping to distinguish it as its own separate field.

INTRODUCTION

Structural engineering is mainly a sub-division of civil engineering where structural engineers are trained to understand, predict, and calculate the stability, strength and rigidity of built structures for buildings and nonbuilding structures, to develop designs and integrate their design with that of other designers, and to supervise construction of projects on site. They can also be involved in the design of machinery, medical equipment, vehicles etc. where structural integrity affects functioning and safety.

Structural engineering theory is based upon applied physical laws and empirical knowledge of the structural performance of different materials and geometries. Structural engineering design utilizes a number of relatively simple structural elements to build complex structural systems. Structural engineers are responsible for making creative and efficient use of funds, structural elements and materials to achieve these goals.

The Eiffel Tower is a historical achievement of structural engineering.

STRUCTURAL ENGINEER (PROFESSIONAL)

Structural engineers are responsible for engineering design and analysis. Entry-level structural engineers may design the individual structural elements of a structure, for example the beams, columns, and floors of a building. More experienced engineers may be responsible for the structural design and integrity of an entire system, such as a building.

Structural engineers often specialize in particular fields, such as bridge engineering, building engineering, pipeline engineering, industrial structures, or special mechanical structures such as vehicles, ships or aircraft.

Structural engineering has existed since humans first started to construct their own structures. It became a more defined and formalised profession with the emergence of the architecture profession as distinct from the engineering profession during the industrial revolution in the late 19th century. Until then, the architect and the structural engineer were usually one and the same - the master builder. Only with the development of specialised knowledge of structural theories that emerged during the 19th and early 20th centuries did the professional structural engineer come into existence.

The role of a structural engineer today involves a significant understanding of both static and dynamic loading, and the structures that are available to resist them. The complexity of modern structures often requires a great deal of creativity from the engineer in order to ensure the structures support and resist the loads they are subjected to. A structural engineer will typically have a four or five year undergraduate degree, followed by a minimum of three years of professional practice before being considered fully qualified. Structural engineers are licensed or accredited by different learned societies and regulatory bodies around the world (for example, the Institution of Structural Engineers in the UK). Depending on the degree course they have studied and/or the jurisdiction they are seeking licensure in, they may be accredited (or licensed) as just structural engineers, or as civil engineers, or as both civil and structural engineers. Another international organisation is IABSE (International Association for Bridge and Structural Engineering). The aim of that association is to exchange knowledge and to advance the practice of structural engineering worldwide in the service of the profession and society.

History of Structural Engineering

Pont du Gard, France, a Roman era aqueduct circa 19 BC.

Structural engineering dates back to 2700 B.C.E. when the step pyramid for PharaohDjoser was built by Imhotep, the first engineer in history known by name. Pyramids were the most common major structures built by ancient civilizations because the structural form of a pyramid is inherently stable and can be almost infinitely scaled (as opposed to most other structural forms, which cannot be linearly increased in size in proportion to increased loads).

However, it's important to note that the structural stability of the pyramid is not primarily a result of its shape. The integrity of the pyramid is intact as long as each of the stones is able to support the weight of the stone above it. The limestone blocks were taken from a quarry near the build site. Since the compressive strength of limestone is anywhere from 30 to 250 MPa (MPa = Pa * 10^6), the blocks will not fail under compression. Therefore, the structural strength of the pyramid stems from the material properties of the stones from which it was built rather than the pyramid's geometry.

Throughout ancient and medieval history most architectural design and construction was carried out by artisans, such as stone masons and carpenters, rising to the role of master builder. No theory of structures existed, and understanding of how structures stood up was extremely limited, and based almost entirely on empirical evidence of 'what had worked before'. Knowledge was retained by guilds and seldom supplanted by advances. Structures were repetitive, and increases in scale were incremental.

No record exists of the first calculations of the strength of structural members or the behavior of structural material, but the profession of structural engineer only really took shape with the Industrial Revolution and the re-invention of concrete (see History of Concrete). The physical sciences underlying structural engineering began to be understood in the Renaissance and have since developed into computer-based applications pioneered in the 1970s.

Timeline

- 1452–1519 Leonardo da Vinci made many contributions
- 1638: Galileo Galilei published the book "Two New Sciences" in which he examined the failure of simple structures

Galileo Galilei published the book "Two New Sciences" in which he examined the failure of simple structures

- 1660: Hooke's law by Robert Hooke
- 1687: Isaac Newton published "Philosophiae Naturalis Principia Mathematica" which contains the Newton's laws of motion

Isaac Newton published "Philosophiae Naturalis Principia Mathematica" which contains the Newton's laws of motion

- 1750: Euler–Bernoulli beam equation
- 1700–1782: Daniel Bernoulli introduced the principle of virtual work
- 1707–1783: Leonhard Euler developed the theory of buckling of columns

Leonhard Euler developed the theory of buckling of columns

- 1826: Claude-Louis Navier published a treatise on the elastic behaviors of structures

- 1873: Carlo Alberto Castigliano presented his dissertation "Intorno ai sistemi elastici", which contains his theorem for computing displacement as partial derivative of the strain energy. This theorem includes the method of *least work* as a special case
- 1874: Otto Mohr formalized the idea of a statically indeterminate structure.
- 1922: Timoshenko corrects the Euler-Bernoulli beam equation
- 1936: Hardy Cross' publication of the moment distribution method, an important innovation in the design of continuous frames.
- 1941: Alexander Hrennikoff solved the discretization of plane elasticity problems using a lattice framework
- 1942: R. Courant divided a domain into finite subregions
- 1956: J. Turner, R. W. Clough, H. C. Martin, and L. J. Topp's paper on the "Stiffness and Deflection of Complex Structures" introduces the name "finite-element method" and is widely recognized as the first comprehensive treatment of the method as it is known today

Structural Failure

The history of structural engineering contains many collapses and failures. Sometimes this is due to obvious negligence, as in the case of the Pétionville school collapse, in which Rev. Fortin Augustin *"constructed the building all by himself, saying he didn't need an engineer as he had good knowledge of construction"* following a partial collapse of the three-story schoolhouse that sent neighbors fleeing. The final collapse killed 94 people, mostly children.

In other cases structural failures require careful study, and the results of these inquiries have resulted in improved practices and greater understanding of the science of structural engineering. Some such studies are the result of forensic engineering investigations where the original engineer seems to have done everything in accordance with the state of the profession and acceptable practice yet a failure still eventuated. A famous case of structural knowledge and practice being advanced in this manner can be found in a series of failures involving box girders which collapsed in Australia during the 1970s.

SPECIALIZATIONS

Building Structures

Structural building engineering includes all structural engineering related to the design of buildings. It is a branch of structural engineering closely affiliated with architecture.

Structural building engineering is primarily driven by the creative manipulation of materials and forms and the underlying mathematical and scientific ideas to achieve an end which fulfills its functional requirements and is structurally safe when subjected to all the loads it could reasonably be expected to experience. This is subtly different from architectural design, which is driven by the creative manipulation of materials and forms, mass, space, volume, texture and light to achieve an end which is aesthetic, functional and often artistic.

Sydney Opera House, designed by Architect Jørn Utzon and structural design by Ove Arup & Partners

Millennium Dome in London, UK, by Richard Rogers and Buro Happold

Burj Khalifa, in Dubai, the world's tallest building, shown under construction in 2007 (since completed)

The architect is usually the lead designer on buildings, with a structural engineer employed as a sub-consultant. The degree to which each discipline actually leads the design depends heavily on the type of structure. Many structures are structurally simple and led by architecture, such as multi-storey office buildings and housing, while other structures, such as tensile structures, shells and gridshells are heavily dependent on their form for their strength, and the engineer may have a more significant influence on the form, and hence much of the aesthetic, than the architect.

The structural design for a building must ensure that the building is able to stand up safely, able to function without excessive deflections or movements which may cause fatigue of structural elements, cracking or failure of fixtures, fittings or partitions, or discomfort for occupants. It must account for movements and forces due to temperature, creep, cracking and imposed loads. It must also ensure that the design is practically buildable within acceptable manufacturing tolerances of the materials. It must allow the architecture to work, and the building services to fit within the building and function (air conditioning, ventilation, smoke extract, electrics, lighting etc.). The structural design of a modern building can be extremely complex, and often requires a large team to complete.

Structural engineering specialties for buildings include:

- Earthquake engineering

- Façade engineering
- Fire engineering
- Roof engineering
- Tower engineering
- Wind engineering

Earthquake Engineering Structures

Earthquake engineering structures are those engineered to withstand earthquakes.

Earthquake-proof pyramid El Castillo, Chichen Itza

The main objectives of earthquake engineering are to understand the interaction of structures with the shaking ground, foresee the consequences of possible earthquakes, and design and construct the structures to perform during an earthquake.

Earthquake-proof structures are not necessarily extremely strong like the El Castillo pyramid at Chichen Itza shown above. In fact, many structures considered strong may in fact be stiff, which can result in poor seismic performance.

One important tool of earthquake engineering is base isolation, which allows the base of a structure to move freely with the ground.

Civil Engineering Structures

Civil structural engineering includes all structural engineering related to the built environment. It includes:

- Bridges
- Dams

- Earthworks
- Foundations
- Offshore structures
- Pipelines
- Power stations
- Railways
- Retaining structures and walls
- Roads
- Tunnels
- Waterways
- reservoir
- Water and wastewater infrastructure

The structural engineer is the lead designer on these structures, and often the sole designer. In the design of structures such as these, structural safety is of paramount importance (in the UK, designs for dams, nuclear power stations and bridges must be signed off by a chartered engineer).

Civil engineering structures are often subjected to very extreme forces, such as large variations in temperature, dynamic loads such as waves or traffic, or high pressures from water or compressed gases. They are also often constructed in corrosive environments, such as at sea, in industrial facilities or below ground.

Mechanical Structures

Car

Motorbike

Principles of structural engineering are applied to variety of mechanical (moveable) structures. The design of static structures assumes they always have the same geometry (in fact, so-called static structures can move significantly, and structural engineering design must take this into account where necessary), but the design of moveable or moving structures must account for fatigue, variation in the method in which load is resisted and significant deflections of structures.

The forces which parts of a machine are subjected to can vary significantly, and can do so at a great rate. The forces which a boat or aircraft are subjected to vary enormously and will do so thousands of times over the structure's lifetime. The structural design must ensure that such structures are able to endure such loading for their entire design life without failing.

These works can require mechanical structural engineering:

- Boilers and pressure vessels
- Coachworks and carriages
- Cranes
- Elevators
- Escalators
- Marine vessels and hulls

Aerospace Structures

An Airbus A380, the world's largest passenger airliner

Design of missile needs in depth understanding of Structural Analysis

Aerospace structure types include launch vehicles, (Atlas, Delta, Titan), missiles (ALCM, Harpoon), Hypersonic vehicles (Space Shuttle), military aircraft (F-16, F-18) and commercial aircraft (Boeing 777, MD-11). Aerospace structures typically consist of thin plates with stiffeners for the external surfaces, bulkheads and frames to support the shape and fasteners such as welds, rivets, screws and bolts to hold the components together.

Nanoscale Structures

A nanostructure is an object of intermediate size between molecular and microscopic (micrometer-sized) structures. In describing nanostructures it is necessary to differentiate between the number of dimensions on the nanoscale. Nanotextured surfaces have one dimension on the nanoscale, i.e., only the thickness of the surface of an object is between 0.1 and 100 nm. Nanotubes have two dimensions on the nanoscale, i.e., the diameter of the tube is between 0.1 and 100 nm; its length could be much greater. Finally, spherical nanoparticles have three dimensions on the nanoscale, i.e., the particle is between 0.1 and 100 nm in each spatial dimension. The terms nanoparticles and ultrafine particles (UFP) often are used synonymously although UFP can reach into the micrometre range. The term 'nanostructure' is often used when referring to magnetic technology.

Structural Engineering for Medical Science

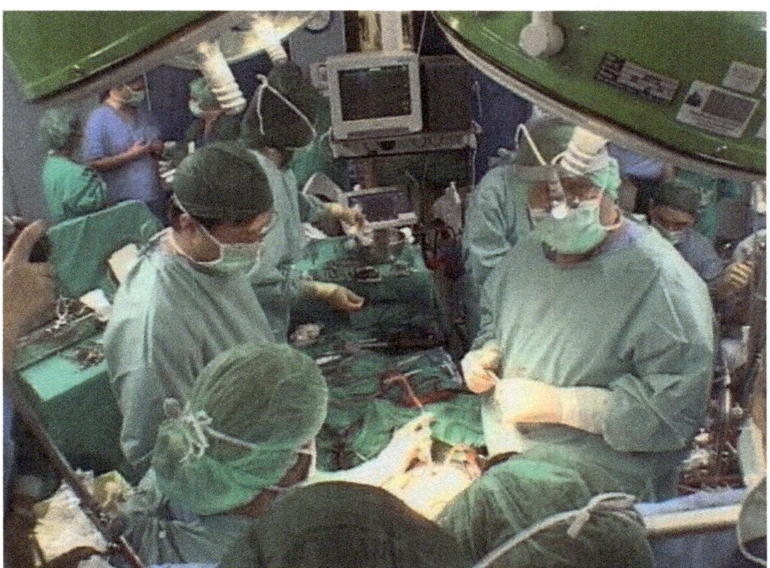

Designing Medical Equipment needs in-depth understanding of Structural Engineering

Medical equipment (also known as armamentarium) is designed to aid in the diagnosis, monitoring or treatment of medical conditions. There are several basic types: Diagnostic equipment includes medical imaging machines, used to aid in diagnosis; equipment includes infusion pumps, medical lasers and LASIK surgical machines; Medical monitors allow medical staff to measure a patient's medical state. Monitors may measure patient vital signs and other parameters including ECG, EEG, blood pressure, and dissolved gases in the blood; Diagnostic Medical Equipment may also be used in the home for certain purposes, e.g. for the control of diabetes mellitus. A biomedical equipment technician (BMET) is a vital component of the healthcare delivery system. Employed primarily by hospitals, BMETs are the people responsible for maintaining a facility's medical equipment.

Structural Elements

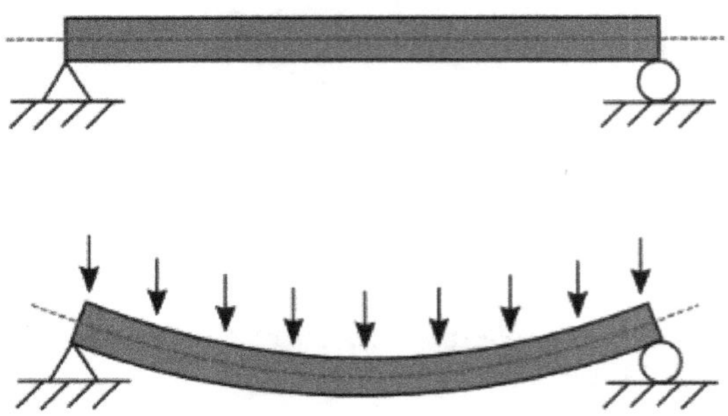

A statically determinate simply supported beam, bending under an evenly distributed load.

Any structure is essentially made up of only a small number of different types of elements:

- Columns
- Beams
- Plates
- Arches
- Shells
- Catenaries

Many of these elements can be classified according to form (straight, plane / curve) and dimensionality (one-dimensional / two-dimensional):

One-dimensional		**Two-dimensional**		
	straight	**curve**	**plane**	**curve**
(predominantly) bending	beam	continuous arch	plate, concrete slab	lamina, dome
(predominant) tensile stress	rope, tie	Catenary	shell	
(predominant) compression	pier, column		Load-bearing wall	

Columns

Columns are elements that carry only axial force - compression - or both axial force and bending (which is technically called a beam-column but practically, just a column). The design of a column must check the axial capacity of the element, and the buckling capacity.

The buckling capacity is the capacity of the element to withstand the propensity to buckle. Its capacity depends upon its geometry, material, and the effective length of the column, which depends upon the restraint conditions at the top and bottom of the column. The effective length is where $K * l$ where l is the real length of the column and K is the factor dependent on the restraint conditions.

The capacity of a column to carry axial load depends on the degree of bending it is subjected to, and vice versa. This is represented on an interaction chart and is a complex non-linear relationship.

Beams

Little Belt: a truss bridge in Denmark

A beam may be defined as an element in which one dimension is much greater than the other two and the applied loads are usually normal to the main axis of the element. Beams and columns are called line elements and are often represented by simple lines in structural modeling.

- cantilevered (supported at one end only with a fixed connection)
- simply supported (supported vertically at each end; horizontally on only one to withstand friction, and able to rotate at the supports)
- fixed (supported at both ends by fixed connection; unable to rotate at the supports)
- continuous (supported by three or more supports)
- a combination of the above (ex. supported at one end and in the middle)

Beams are elements which carry pure bending only. Bending causes one part of the section of a beam (divided along its length) to go into compression and the other part into tension. The compression part must be designed to resist buckling and crushing, while the tension part must be able to adequately resist the tension.

Trusses

The McDonnell Planetarium by Gyo Obata in St Louis, Missouri, USA, a concrete shell structure

The 630 foot (192 m) high, stainless-clad (type 304) Gateway Arch in Saint Louis, Missouri

A truss is a structure comprising members and connection points or nodes. When members are connected at nodes and forces are applied at nodes members can act in tension or in compression. Members acting in compression are referred to as compression members or struts while members acting in tension are referred to as tension members or ties. Most trusses use gusset plates to connect intersecting elements. Gusset plates are relatively flexible and unable to transfer bending moments at the connections, thus allowing the truss members to act primarily in tension or compression.

Trusses are usually utilised in large-span structures, where it would be uneconomical to use solid beams.

Plates

Plates carry bending in two directions. A concrete flat slab is an example of a plate. Plates are understood by using continuum mechanics, but due to the complexity involved they are most often designed using a codified empirical approach, or computer analysis.

They can also be designed with yield line theory, where an assumed collapse mechanism is analysed to give an upper bound on the collapse load (see Plasticity). This technique is used in practice but because the method provides an upper-bound, i.e. an unsafe prediction of the collapse load, for poorly conceived collapse mechanisms great care is needed to ensure that the assumed collapse mechanism is realistic.

Shells

Shells derive their strength from their form, and carry forces in compression in two directions. A dome is an example of a shell. They can be designed by making a hanging-chain model, which will act as a catenary in pure tension, and inverting the form to achieve pure compression.

Arches

Arches carry forces in compression in one direction only, which is why it is appropriate to build arches out of masonry. They are designed by ensuring that the line of thrust of the force remains within the depth of the arch. It is mainly used to increase the bountifulness of any structure.

Catenaries

Catenaries derive their strength from their form, and carry transverse forces in pure tension by deflecting (just as a tightrope will sag when someone walks on it). They are almost always cable or fabric structures. A fabric structure acts as a catenary in two directions.

Structural Engineering Theory

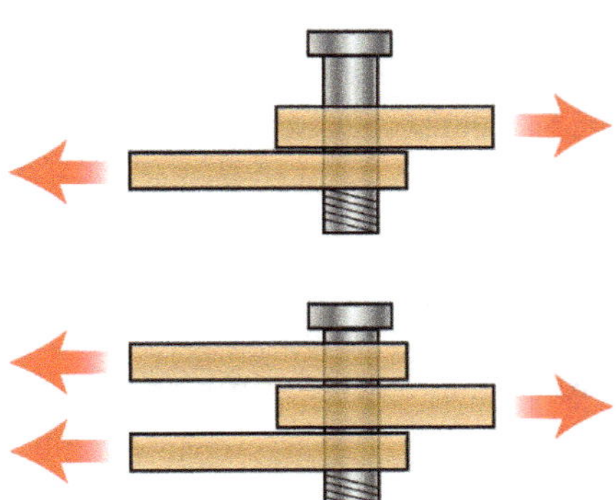

Figure of a bolt in shear stress. Top figure illustrates single shear, bottom figure illustrates double shear.

Structural engineering depends upon a detailed knowledge of applied mechanics, materials science and applied mathematics to understand and predict how structures support and resist self-weight and imposed loads. To apply the knowledge successfully a structural engineer generally requires detailed knowledge of relevant empirical and theoretical design codes, the techniques of structural analysis, as well as some knowledge of the corrosion resistance of the materials and structures, especially when those structures are exposed to the external environment. Since the 1990s, specialist software has become available to aid in the design of structures, with the functionality to assist in the drawing, analyzing and designing of structures with maximum precision; examples include AutoCAD, StaadPro, ETABS, Prokon, Revit Structure, Inducta RCB, etc. Such software may also take into consideration environmental loads, such as from earthquakes and winds.

Materials

Structural engineering depends on the knowledge of materials and their properties, in order to understand how different materials support and resist loads.

Common structural materials are:

- Iron: Wrought iron, Cast iron
- Concrete: Reinforced concrete, Prestressed concrete
- Alloy: Steel, Stainless steel
- Masonry
- Timber: Hardwood, Softwood
- Aluminium
- Composite materials: Plywood
- Other structural materials:Adobe, Bamboo, Carbon fibre, Fiber reinforced plastic, Mudbrick, Roofing materials

STRUCTURAL ELEMENT

Structural elements are used in structural analysis to split a complex structure into simple elements. Within a structure, an element cannot be broken down (decomposed) into parts of different kinds (e.g., beam or column).

Structural elements can be linear, surfaces or volumes.

Linear elements:

- Rod - axial loads
- Beam - axial and bending loads
- Struts or Compression members- compressive loads
- Ties, Tie rods, eyebars, guy-wires, suspension cables, or wire ropes - tension loads

Surface elements:

- membrane - in-plane loads only
- shell - in plane and bending moments
- Concrete slab
  - deck
- shear panel - shear loads only

Volumes:

- Axial, shear and bending loads for all three dimensions

Cylinder

In its simplest form, a **cylinder** (from Greek κύλινδρος – *kulindros*, "roller, tumbler") is the surface formed by the points at a fixed distance from a given straight line called the **axis** of the cylinder. It is one of the most basic curvilinear geometric shapes.

Common Use

Commonly the word *cylinder* is understood to refer to a finite section of a ***right circular cylinder*** having a finite height with circular ends perpendicular to the axis as shown in the figure. If the ends are open, it is called an open cylinder. If the ends are closed by flat surfaces it is called a solid cylinder. The formulae for the surface area and the volume of such a cylinder have been known since deep antiquity.

Volume

If the cylinder has a radius r and length (height) h, then its volume is given by

$$V = \pi r^2 h$$

Having a right circular cylinder with a height h units and a base of radius r units with the coordinate axes chosen so that the origin is at the center of one base and the height is measured along the positive x-axis. A plane section at a distance of x units from the origin has an area of $A(x)$ square units where

$$A(x) = \pi r^2$$

or

$$A(y) = \pi r^2$$

An element of volume, is a right cylinder of base area Aw_i square units and a thickness of $\Delta_i x$ units. Thus if V cubic units is the volume of the right circular cylinder, by Riemann sums,

$$\text{Volume of cylinder} = \lim_{||\Delta \to 0||} \sum_{i=1}^{n} A(w_i)\Delta_i x$$
$$= \int_0^h A(y)\, dy$$
$$= \int_0^h \pi r^2\, dy$$
$$= \pi r^2 h$$

Using cylindrical coordinates, the volume can be calculated by integration over

$$= \int_0^h \int_0^{2\pi} \int_0^r s\, ds\, d\phi\, dz$$
$$= \pi r^2 h$$

Surface Area

Still using a radius r and length (height) h, the surface area of a cylinder is made up of three parts:

- the area of the top: πr^2
- the area of the bottom: πr^2
- the area of the side: $2\pi rh$

The area of the top and bottom is always the same, and is also called the base area, B. The area of the side is also known as the lateral area, L.

An open cylinder does not include either top or bottom elements, and therefore has surface area (lateral area)

$L = 2\pi rh.$

The surface area of a closed cylinder is made up the sum of all three components: top, bottom and side. Its surface area is

$A = 2\pi r^2 + 2\pi rh = 2\pi r(r + h) = \pi d(r + h) = L + 2B,$

where d is the diameter.

For a given volume, the closed cylinder with the smallest surface area has $h = 2r$. Equivalently, for a given surface area, the closed cylinder with the largest volume has $h = 2r$, i.e. the cylinder fits snugly in a cube (height = diameter).

Cylindric Sections

Cylindric sections are the intersections of cylinders with planes. For a right circular cylinder, there are four possibilities. A plane tangent to the cylinder meets the cylinder in a single straight line segment. Moved while parallel to itself, the plane either does not intersect the cylinder or intersects it in two parallel line segments. All other planes intersect the cylinder in an ellipse or, when they are perpendicular to the axis of the cylinder, in a circle.

Eccentricity e of the cylindric section and semi-major axis a of the cylindric section depend on the radius of the cylinder r and the angle between the secant plane and cylinder axis α in the following way:

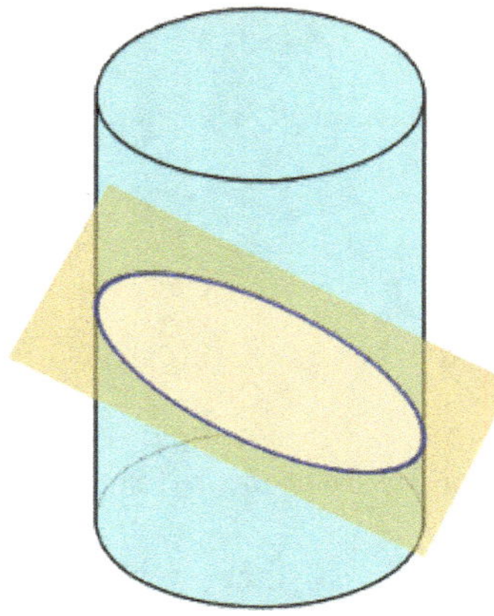

Cylindric section.

Tycho Brahe Planetarium building, Copenhagen, its roof being an example of a cylindric section

$$e = \cos\alpha$$

$$a = \frac{r}{\sin\alpha}$$

Other Types of Cylinders

In differential geometry, a cylinder is defined more broadly as any ruled surface spanned by a one-parameter family of parallel lines. A cylinder whose cross section is an ellipse, parabola, or hyperbola is called an elliptic cylinder, parabolic cylinder, or hyperbolic cylinder respectively.

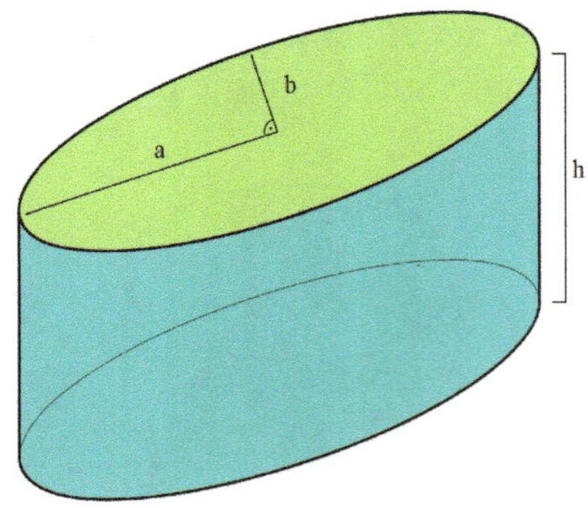

An elliptic cylinder with the half-axes *a* and *b* for the surface ellipse and the height *h*.

An elliptic cylinder is a quadric surface, with the following equation in Cartesian coordinates:

$$\left(\frac{x}{a}\right)^2+\left(\frac{y}{b}\right)^2=1$$

This equation is for an elliptic cylinder, a generalization of the ordinary, circular cylinder ($a = b$). Elliptic cylinders are also known as cylindroids, but that name is ambiguous, as it can also refer to the Plücker conoid. The volume of an elliptic cylinder with height h is $V=\int_0^h A(x)dx=\int_0^h \pi ab dx=\pi ab\int_0^h dx=\pi abh.$ Even more general than the elliptic cylinder is the **generalized cylinder**: the cross-section can be any curve.

The cylinder is a *degenerate quadric* because at least one of the coordinates (in this case z) does not appear in the equation.

An **oblique cylinder** has the top and bottom surfaces displaced from one another.

There are other more unusual types of cylinders. These are the *imaginary elliptic cylinders*:

$$\left(\frac{x}{a}\right)^2+\left(\frac{y}{b}\right)^2=-1$$

the *hyperbolic cylinder*:

$$\left(\frac{x}{a}\right)^2-\left(\frac{y}{b}\right)^2=1$$

and the *parabolic cylinder*:

$$x^2+2ay=0$$

Right Circular Hollow Cylinder

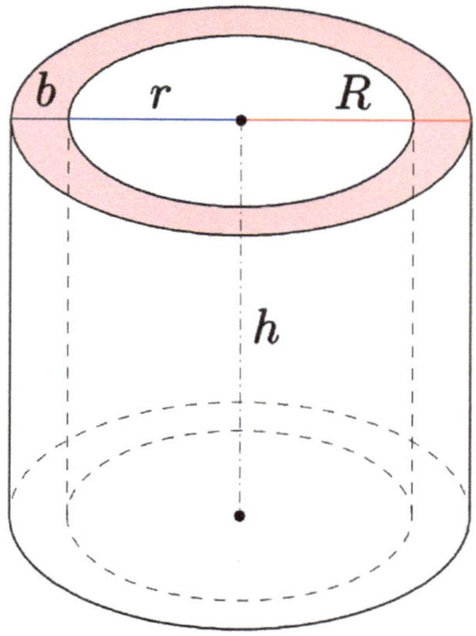

Hollow cylinder(geometry)

A right circular hollow cylinder is a three-dimensional region bounded by two concentric cylindrical sections and two parallel annular bases perpendicular to the cylinder's axis.

Let the height be h, internal radius r, and external radius R. The volume is given by

$$V = \pi(R^2 - r^2)h.$$

The surface area is given by

$$A = 2\pi(R + r)h + 2\pi(R^2 - r^2).$$

About an Arbitrary Axis

Consider an infinite cylinder whose axis lies along the vector

$$\vec{v} = (\alpha, \beta, \gamma)$$

We make use of spherical coordinates:

$$\rho^2 = \alpha^2 + \beta^2 + \gamma^2$$

$$\theta = \arctan\left(\frac{\beta}{\alpha}\right)$$

$$\phi = \arcsin\left(\frac{\gamma}{\rho}\right)$$

These variables can be used to define A and B, the orthogonal vectors that form the basis for the cylinder:

$$A = -x\sin(\theta) + y\cos(\theta)$$

$$B = -x\cos(\theta)\sin(\phi) + y\sin(\theta)\cos(\phi) + z\cos(\phi)$$

With these defined, we may use the familiar formula for a cylinder:

$$A^2 + B^2 = R^2$$

where R is the radius of the cylinder. These results are usually derived using rotation matrices.

Projective Geometry

In projective geometry, a cylinder is simply a cone whose apex is at infinity, which corresponds visually to a cylinder in perspective appearing to be a cone towards the sky.

In projective geometry, a cylinder is simply a cone whose apex is at infinity.

This is useful in the definition of degenerate conics, which require considering the cylindrical conics.

Related Polyhedra

A *cylinder* can be seen as a polyhedral limiting case of an n-gonal prism where *n* approaches infinity. It can also be seen as a dual of a bicone as an infinite-sided bipyramid.

Family of uniform prisms

Polyhedron										
Coxeter		4	5	6	7	8	9	10	11	12
Tiling										
Config.	3.4.4	4.4.4	5.4.4	6.4.4	7.4.4	8.4.4	9.4.4	10.4.4	11.4.4	12.4.4

BEAM (STRUCTURE)

A **beam** is a structural element that is capable of withstanding load primarily by resisting against bending. The bending force induced into the material of the beam as a result of the external loads, own weight, span and external reactions to these loads is called a bending moment. Beams are characterized by their profile (shape of cross-section), their length, and their material.

Beams are traditionally descriptions of building or civil engineering structural elements, but smaller structures such as truck or automobile frames, machine frames, and other mechanical or structural systems contain beam structures that are designed and analyzed in a similar fashion.

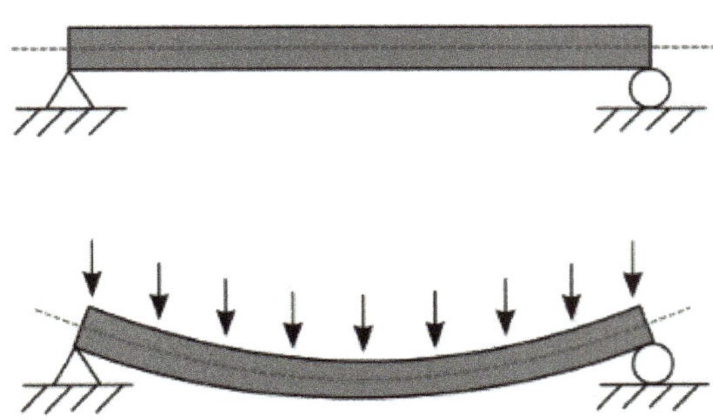

A statically determinate beam, bending (sagging) under a uniformly distributed load

Historically beams were squared timbers but are also metal, stone, or combinations of wood and metal such as a flitch beam. Beams generally carry vertical gravitational forces but can also be used to carry horizontal loads (e.g., loads due to an earthquake or wind or in tension to resist rafter thrust as a tie beam or (usually) compression as a collar beam). The loads carried by a beam are transferred to columns, walls, or girders, which then transfer the force to adjacent structural compression members. In light frame construction joists may rest on beams.

In carpentry a beam is called a plate as in a sill plate or wall plate, beam as in a summer beam or dragon beam.

Classification of Beams Based on Supports

In engineering, beams are of several types:

- Simply supported - a beam supported on the ends which are free to rotate and have no moment resistance.
- Fixed - a beam supported on both ends and restrained from rotation.
- Over hanging - a simple beam extending beyond its support on one end.
- Double overhanging - a simple beam with both ends extending beyond its supports on both ends.
- Continuous - a beam extending over more than two supports.
- Cantilever - a projecting beam fixed only at one end.
- Trussed - a beam strengthened by adding a cable or rod to form a truss.

Area Moment of Inertia

In the beam equation I is used to represent the second moment of area. It is commonly known as the moment of inertia, and is the sum, about the neutral axis, of dA*r^2, where r is the distance from the neutral axis, and dA is a small patch of area. Therefore, it encompasses not just how much area the beam section has overall, but how far each bit of area is from the axis, squared. The greater I is, the stiffer the beam in bending, for a given material.

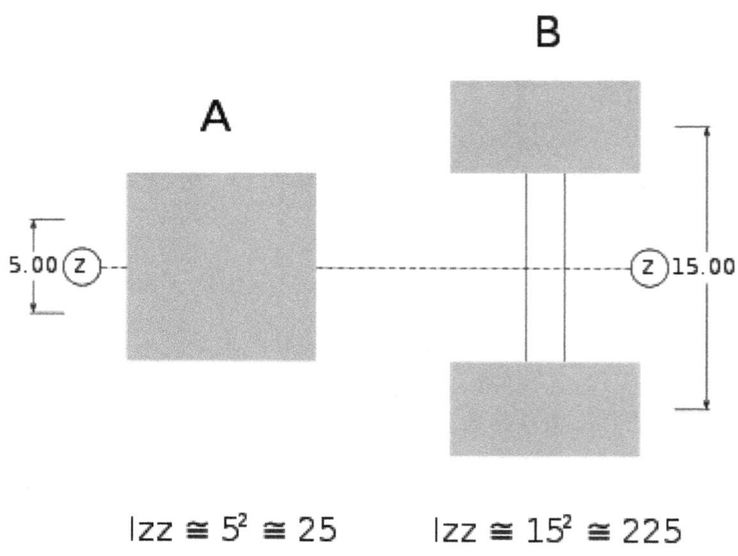

Diagram of stiffness of a simple square beam (A) and universal beam (B). The universal beam flange sections are three times further apart than the solid beam's upper and lower halves. The second moment of inertia of the universal beam is nine times that of the square beam of equal cross section (universal beam web ignored for simplification)

Stress in Beams

Internally, beams experience compressive, tensile and shear stresses as a result of the loads applied to them. Typically, under gravity loads, the original length of the beam is slightly reduced to enclose a smaller radius arc at the top of the beam, resulting in compression, while the same original beam length at the bottom of the beam is slightly stretched to enclose a larger radius arc, and so is under tension. The same original length of the middle of the beam, generally halfway between the top and bottom, is the same as the radial arc of bending, and so it is under neither compression nor tension, and defines the neutral axis (dotted line in the beam figure). Above the supports, the beam is exposed to shear stress. There are some reinforced concrete beams in which the concrete is entirely in compression with tensile forces taken by steel tendons. These beams are known as prestressed concrete beams, and are fabricated to produce a compression more than the expected tension under loading conditions. High strength steel tendons are stretched while the beam is cast over them. Then, when the concrete has cured, the tendons are slowly released and the beam is immediately under eccentric axial loads. This eccentric loading creates an internal moment, and, in turn, increases the moment carrying capacity of the beam. They are commonly used on highway bridges.

The primary tool for structural analysis of beams is the Euler–Bernoulli beam equation. Other mathematical methods for determining the deflection of beams include "method of virtual work" and the "slope deflection method". Engineers are interested in determining deflections because the beam may be in direct contact with a brittle material such as glass. Beam deflections are also minimized for aesthetic reasons. A visibly sagging beam, even if structurally safe, is unsightly and to be avoided. A stiffer beam (high modulus of elasticity and high second moment of area) creates less deflection.

Mathematical methods for determining the beam forces (internal forces of the beam and the forces that are imposed on the beam support) include the "moment distribution method", the force or flexibility method and the direct stiffness method.

General Shapes

Most beams in reinforced concrete buildings have rectangular cross sections, but a more efficient cross section for a beam is an I or H section which is typically seen in steel construction. Because of the parallel axis theorem and the fact that most of the material is away from the neutral axis, the second moment of area of the beam increases, which in turn increases the stiffness.

An I-beam is only the most efficient shape in one direction of bending: up and down looking at the profile as an I. If the beam is bent side to side, it functions as an H where it is less efficient. The most efficient shape for both directions in 2D is a box (a square shell) however the most efficient shape for bending in any direction is a cylindrical shell or tube. But, for unidirectional bending, the I or wide flange beam is superior.

Efficiency means that for the same cross sectional area (volume of beam per length) subjected to the same loading conditions, the beam deflects less.

Other shapes, like L (angles), C (channels) or tubes, are also used in construction when there are special requirements.

Thin Walled Beams

A thin walled beam is a very useful type of beam (structure). The cross section of *thin walled beams* is made up from thin panels connected among themselves to create closed or open cross sections of a beam (structure). Typical closed sections include round, square, and rectangular tubes. Open sections include I-beams, T-beams, L-beams, and so on. Thin walled beams exist because their bending stiffness per unit cross sectional area is much higher than that for solid cross sections such a rod or bar. In this way, stiff beams can be achieved with minimum weight. Thin walled beams are particularly useful when the material is a composite laminates. Pioneer work on composite laminates thin walled beams was done by Librescu.

STRUT

A **strut** is a structural component designed to resist longitudinal compression. Struts provide outwards-facing support in their lengthwise direction, which can be used to keep two other components separate, performing the opposite function of a *tie*. They are commonly used in architecture and engineering.

Struts on the undercarriage, wings and tailplane of an Antonov An-2 biplane

Automobiles

As components of an automobile chassis, struts can be passive braces to reinforce the chassis and/or body, or active components of the suspension. An automotive suspension strut combines the primary function of a shock absorber (as a damper), with the ability to support sideways loads not along its axis of compression, somewhat similar to a sliding pillar suspension, thus eliminating the need for an upper suspension arm. This means that a strut must have a more rugged design, with mounting points near its middle for attachment of such loads.

The most common form of strut in an automobile is the MacPherson strut. MacPherson struts are often purchased by the automakers in sets of four completed sub-assemblies: These can be mounted on the car bodies as part of the manufacturers' own assembly operations. The MacPherson strut combines a shock absorber and a spring in a single unit, by means of which each wheel is attached to the car body.

Transportation-related struts are used in "load bearing" applications ranging from both highway and off-road suspensions to automobile hood and hatch window supports to aircraft wing supports. In vehicle suspension systems, struts are most commonly an assembly of coil-over spring and shock absorber. Other variants to using a coil-over spring as the compressible load bearer include support via pressurized nitrogen gas acting as the spring, and rigid (hard tail) support which provides neither longitudinal compression/ extension nor damping.

Strut Failure

Signs of strut failure include a popping or clicking noise when the steering wheel of the vehicle is turned, especially when the vehicle is not in motion. This noise is caused by a worn out bearing. Most strut failure is gradual, and becomes evident as a more "trampoline"-like ride, rough travel over bumps, and tire noise/ vibrations. This wear affects all of the strut components in a feedback loop: the more worn out the shock absorbers become, the more worn out the spring will become, the more worn out the mount will become, and the more this will affect the shock absorbers. A tendency for the front of the vehicle to "nose dive" when applying the brakes and to rear upwards ("squatting") when accelerating are other signs of strut failure, as is difficulty handling the vehicle generally. These symptoms will only tend to worsen until the struts are replaced with new components (struts cannot be repaired when worn out, but must be removed entirely). Many strut "replacements" on non-commercial vehicles are in fact only replacements of the shock absorber and not the other components.

Vehicle handling is one of the easiest aspects of a potential new car purchase for a buyer to assess without professional assistance. A car which nose dives when stopped suddenly, squats when rapidly accelerated, or tends to "roll" when going around curves in the roadway are all signs that the vehicle will likely require strut replacement in the near future, regardless of other aspects of the vehicle such as the apparent condition of its body or its interior.

Architecture and Construction

In piping, struts restrain movement of a component in one direction while allowing movement or contraction in another direction. Strut channel made from steel, aluminium, or fibre-reinforced plastic is used heavily in the building industry and is often used in the support of cable trays and other forms of cable management, and pipes support systems.

Strut is a common name in timber framing for a support or brace of scantlings lighter than a post. Frequently struts are found in roof framing from either a tie beam or a king post to a principal rafter. Struts may be vertically plumb or leaning (then called canted, raking, or angled) and may be straight or curved. In the U.K., strut is generally used in a sense of a lighter duty piece: a king post carries a ridge beam but a king strut does not, a queen post carries a plate but a queen strut does not, a crown post carries a crown plate but a crown strut does not.

Strutting or blocking between floor joists adds strength to the floor system.

Aircraft

Bracing struts and wires of many kinds were extensively used in early aircraft to stiffen and strengthen, and sometimes even to form, the main functional airframe. Throughout the 1920s and 1930s they fell out of use in favour of the low-drag cantilever construction. Most aircraft bracing struts are principally loaded in compression, with wires taking the tension loads. Lift struts came into increasing use during the changeover period and remain in use on smaller aircraft today where ultimate performance is not an issue. Typically they are applied to a high-wing monoplane and act in tension during flight.

Struts have also been widely used for purely structural reasons to attach engines, landing gear and other loads. The oil-sprung legs of retractable landing gear are still called *Oleo struts*.

Human Anatomy

Part of the functionality of the clavicle is to serve as a strut between the scapula and sternum, resisting forces that would otherwise bring the upper limb close to the thorax. Keeping the upper limb away from the thorax is vital for its range of motion. Complete lack of clavicles may be seen in cleidocranial dysostosis, and the abnormal proximity of the shoulders to the median plane exemplifies the clavicles' importance as struts.

Gallery

ORI Strut, Off-road application of strut

Strut bar

Compression strut on Piper Pawnee (low-wing aircraft)

Tension strut on Shorts 360 (high-wing aircraft)

MacPherson Strut suspension

Nosewheel oleo strut on Su-30MKI aircraft

TIE (ENGINEERING)

A tie, strap, tie rod, eyebar, guy-wire, suspension cables, or wire ropes, are an examples of linear structural components designed to resist tension. It is the opposite of a strut or column, which is designed to resist compression. Ties may be made of any tension resisting material.

A hurricane tie used to fasten a rafter to a stud

Application in wood construction

In wood frame construction they are generally made of galvanized steel. Wood framing ties generally have holes allowing them to be fastened to the wood structure by nails or screws. The number and type of nails are specific to the tie and its use. The manufacturer generally specifies information as to the connection method for each of their products. Among the most common wood framing ties used is the hurricane tie or seismic tie used in the framing of wooden structures where wind uplift or seismic overturning is a concern.

Hurricane Tie

Hurricane ties are in place at the top of the wall as the roof trusses are being placed.

A hurricane tie is used to help make a structure (specifically wooden structures) more resistant to high winds (such as in hurricanes), resisting uplift, racking, overturning, and sliding. Each of the crucial connections in a structure, that would otherwise fail under the pressures of high winds, have a corresponding type of tie, generally made of galvanized or stainless steel, and intended to resist hurricane-force and other strong winds.

A connecting tie that provides a continuous structural load transfer path from the top of a building to its foundation, helping to protect buildings from damage resulting from high wind. These devices are primarily used in areas affected by high winds including hurricanes and are generally suitable for any area that may be impacted by windstorm damage. They are also known as hurricane clip or strips. Among the most common style used along the gulf coast area are plywood fasteners or oriented strand boards over the windows and openings of brick homes. Hurricane clips meet the minimum requirements for code approval and are only as strong as their weakest install point; which is generally ½ the strength of the wood.

A hurricane clip has two meanings in building construction:(1) A connecting tie that provides a continuous structural load transfer path from the top of a building to its foundation, helping to protect buildings from damage resulting from high wind. These devices are primarily used in areas affected by high winds including hurricanes and are generally suitable for any area that may be impacted by windstorm damage. They are also known as hurricane *ties* or *strips*;(2) A tie which is attached to roof tiles to keep them from blowing off a roof. These devices are also known as *wind clips* and *hurricane side clips*

Seismic Tie

Seismic tie provides facility to securely fix cabinets, bookcases, desks, appliances, machinery & equipment to walls and/or floors to constrain their movement during earthquakes.

Girder Tiedown

Top mount, face mount, sloped/skewed, and variable pitch hangers for dimensional lumber, engineered wood I-joists, structural composite lumber and masonry wall. To give added strength in increase various load requirements over wood only.

Strap Tie

When building subfloor the joists must always bear on the ledge for all it support. The use of steel stap tie to connect opposite joist when the top of the joists and beam are flush.

Twist Strap

Twist straps provide a tension connection between two wood members. They resist uplift at the heel of a truss economically.When the strengthening is being done from the inside, the ideal connector to use is one that connects rafters or trusses directly to wall studs. This can only be done where the rafter or trusses are immediately above or immediately to the side of studs below. In that case a twist strap connector can be used.

Floor Span Connector

A connector for connecting wall studs of two adjacent floors in a light frame building structure, the connector having a first attachment tab, a seat member, a diagonally slanted support leg, and a second attachment tab, all substantially planar. The connector is intended to be paired and the paired connectors joined by an elongated tie member that pierces the sill plates of the intervening floor structure.

Angle Tie

Sometimes referred to as an angle brace. The Angle tie is used to prevent displacement of building elements due to thrust.A brace/tie across an interior angle of a wooden frame, forming the hypotenuse and securing the two side pieces together.

Z-Clip

Similar to a French Cleat, A Z-Clips allow you to install wall panels without screwing into the front of the panels. The clips provide a secure mount for wall panels, partitions, frames, cabinets, and more. Once installed, clips wedge together to lock panels in place. To disengage panels, simply lift and remove.

Plate

Metal connector plates.

Rafter Tie (and Tie-Beams)

Rafter ties are designed to tie together the bottoms of opposing rafters on a roof, to resist the outward thrust where the roof meets the house ceiling and walls. This helps keep walls from spreading due to the weight of the roof and anything on it, notably wet snow. In many or most homes, the ceiling joists also serve as the rafter ties. When the walls spread, the roof ridge will sag. A sagging ridge is one clue that the home may lack adequate rafter ties. Rafter ties form the bottom chord of a simple triangular roof truss. They resist the out-thrust of a triangle that's trying to flatten under the roof's own weight or snow load. They are placed in the bottom one-third of the roof height. Rafter ties are always required unless the roof has a structural (self-supporting) ridge, or is built using engineered trusses. A lack of rafter ties is a serious structural issue in a conventionally framed roof.

The 15th-century tie-beam roof at St Marys Church, Radnage, Buckinghamshire in England

A wooden beam serving this purpose is known as a tie-beam and a roof incorporating tie-beams is known as a tie-beam roof.

THIN-SHELL STRUCTURE

Thin-shell structures are also called plate and shell structures. They are lightweight constructions using shell elements. These elements, typically curved, are assembled to make large structures. Typical applications include aircraft fuselages, boat hulls, and the roofs of large buildings.

Shell structure of the TWA Flight Center Building by Eero Saarinen, John F. Kennedy International Airport, New York

The Forest Opera, an open-air amphitheatre in Sopot, Poland, with a membrane roof.

Definition

A thin shell is defined as a shell with a thickness which is small compared to its other dimensions and in which deformations are not large compared to thickness. A primary difference between a shell structure and a plate structure is that, in the unstressed state, the shell structure has curvature as opposed to the plates structure which is flat. Membrane action in a shell is primarily caused by in-plane forces (plane stress), but there may be secondary forces resulting from flexural deformations. Where a flat plate acts similar to a beam with bending and shear stresses, shells are analogous to a cable which resists loads through tensile stresses. The ideal thin shell must be capable of developing both tension and compression.

Types

The most popular types of thin-shell structures are:

- Concrete shell structures, often cast as a monolithic dome or stressed ribbon bridge or saddle roof
- Lattice shell structures, also called gridshell structures, often in the form of a geodesic dome or a hyperboloid structure
- Membrane structures, which include fabric structures and other tensile structures, cable domes, and pneumatic structures.

Great Court, with a lattice thin-shell roof by Buro Happold with Norman Foster, British Museum, London

CONCRETE SLAB

A concrete slab is a common structural element of modern buildings. Horizontal slabs of steel reinforced concrete, typically between 4 and 20 inches (100 and 500 millimeters) thick, are most often used to construct floors and ceilings, while thinner slabs are also used for exterior paving. Sometimes these thinner slabs, ranging from 2 inches (51 mm) to 6 inches (150 mm) thick, are called mud slabs, particularly when used under the main floor slabs or in crawl spaces.

In many domestic and industrial buildings a thick concrete slab, supported on foundations or directly on the subsoil, is used to construct the ground floor of a building. These can either be "ground-bearing" or "suspended" slabs. In high rise buildings and skyscrapers, thinner, pre-cast concrete slabs are slung between the steel frames to form the floors and ceilings on each level.

On the technical drawings, reinforced concrete slabs are often abbreviated to "r.c.c.slab" or simply "r.c.".

Suspended slab under construction, with the formwork still in place.

Suspended slab formwork and rebar in place, ready for concrete pour. On reinforced concrete blockwork supporting walls.

Thermal Performance

There are two main thermal considerations. The first is the question of insulating a floor slab. In older buildings, concrete slabs cast directly on the ground can drain heat from a room. In modern construction, concrete slabs are usually cast above a layer of insulation such as expanded polystyrene, and the slab may contain underfloor heating pipes. However, there are still uses for an uninsulated slab, typically in outbuildings which are not heated or cooled to room temperature. In those cases, casting the slab directly onto a rocky substrate will maintain the slab at or near the temperature of the substrate throughout the year, and can prevent both freezing and overheating.

The second consideration is the high thermal mass, which applies to walls and floors, or wherever the concrete is used within the thermal envelope. It is a disadvantage where the rooms are heated intermittently and require a quick response, as the concrete takes time to warm up, causing a delay in warming the building. But it is an advantage in climates with large daily temperature swings, where the slab acts as a regulator, keeping the building cool by day and warm by night.

Design

For a suspended slab, there are a number of designs to improve the strength-to-weight ratio. In all cases the top surface remains flat, and the underside is modulated:

- *Corrugated*, usually where the concrete is poured into a corrugated steel tray. This improves strength and prevents the slab from bending under its own weight. The corrugations run across the short dimension, from side to side.
- A *ribbed slab*, giving considerable extra strength on one direction.
- A *waffle slab*, giving added strength in both directions and it looks hollow from bottom .

Reinforcement design

- A *one-way slab* needs moment resisting reinforcement only in its short-direction because the moment along long axes is so small that it can be neglected. When the ratio of the length of long direction to short direction of a slab is greater than 2 it can be considered as a one way slab.

Long direction = ly Short direction = lx

One way slab IF ly/lx > 2

- A *two-way slab* needs moment resisting reinforcement in both directions. If the ratio of the lengths of long and short side is less than two then moment in both direction should be considered in design.

Two way slab IF ly/lx < 2

Construction

A concrete slab may be prefabricated or on site. Prefabricated concrete slabs are built in a factory and transported to the site, ready to be lowered into place between steel or concrete beams. They may be pre-stressed (in the factory), post-stressed (on site), or unstressed. It is vital that the wall supporting structure is built to the correct dimensions, or the slabs may not fit.

In site concrete slabs are built on the building site using formwork - a type of boxing into which the wet concrete is poured. If the slab is to be reinforced, the rebars are positioned within the formwork before the concrete is poured in. Plastic tipped metal, or plastic bar chairs are used to hold the rebar away from the bottom and sides of the form-work, so that when the concrete sets it completely envelops the reinforcement. For a *ground slab*, the form-work may consist only of sidewalls pushed into the ground. For a *suspended slab*, the form-work is shaped like a tray, often supported by a temporary scaffold until the concrete sets.

The formwork is commonly built from wooden planks and boards, plastic, or steel. On commercial building sites today, plastic and steel are more common as they save labour. On low-budget sites, for instance when laying a concrete garden path, wooden planks are very common. After the concrete has set the wood may be removed, or left there permanently.

In some cases formwork is not necessary - for instance, a ground slab surrounded by brick or block foundation walls, where the walls act as the sides of the tray and hardcore acts as the base.

DECK (BRIDGE)

A **bridge deck** or **road bed** is the roadway, or the pedestrian walkway, surface of a bridge, and is one structural element of the superstructure of a bridge. It is not to be confused with any deck of a ship. The deck may be constructed of concrete, steel, open grating, or wood. Sometimes the deck is covered with asphalt concrete or other pavement. The concrete deck may be an integral part of the bridge structure (T-beam or double tee structure) or it may be supported with I-beams or steel girders.

When a bridge deck is installed in a through truss, it is sometimes called a floor system. A **suspended bridge deck** will be suspended from the main structural elements on a suspension or arch bridge. On some bridges, such as a tied-arch or a cable-stayed, the deck is a primary structural element, carrying tension or compression to support the span.

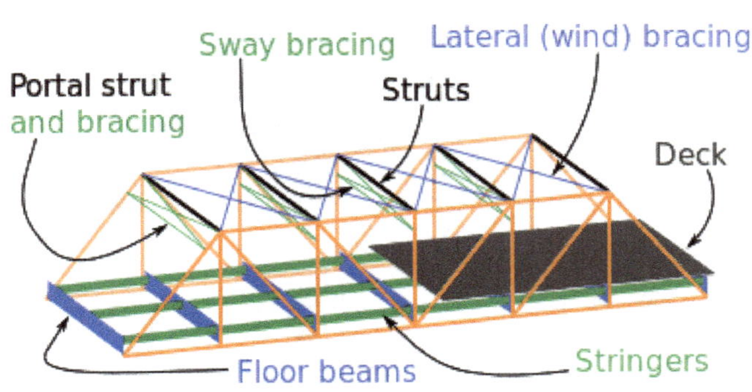

The various part of a truss bridge.

Structural Analysis

Structural engineers have several principal categories of bridge decks, for the purposes of analytic techniques. A *beam deck* is one where the deck and any supporting structure act together as a single beam. A *grid deck* uses beams and diaphragms as the supporting structure. The supporting system of a grid deck is analyzed using a grillage analysis. A *slab deck* is one where the deck is analyzed as a plate. If the slab has a stiffness that is different in two directions (at right angles), then the deck is known and analyzed as an orthotropic deck. A *beam and slab deck* is one where the beams may deflect somewhat independently, and any transverse forces are carried in the deck. A *cellular deck* is one where a number of thin slabs and webs will enclose cells within the deck. A *boxgirder deck* is one where the deck forms the top of the box girder during analysis.

Railway Bridge Decks

An open-deck railway bridge in Leflore County, Mississippi

In the case of a railway bridge, when the railway track and ties are supported on load carrying elements of the superstructure (floor beams, stringers or girders), it is called an *open deck*. When the track rests upon ballast, which is then carried by the superstructure of the bridge, it is called a *ballasted deck*. A third type of rail support is *direct fixation*, where ties are not used and the rails are anchored directly to the superstructure of the bridge.

STRUCTURAL SYSTEM

The term **structural system** or **structural frame** in structural engineering refers to load-resisting sub-system of a structure. The structural system transfers loads through interconnected structural components or members.

Commonly used structures can be classified into five major categories, depending on the type of primary stress that may arise in the members of the structures under major design loads. However any two or more of the basic structural types described in the following may be combined in a single structure, such as a building or a bridge in order to meet the structures functional requirements.

- Tensile structures: Members of tensile structures are subjects to pure tension under the action of external loads. Because the tensile stress is uniformly distributed over the cross-sectional area of members, the material of such a structure is utilized in the most efficient manner.
- Compressive structures: Compression structures develop mainly compressive stresses under the action of axial loads. Because compressive structures are susceptible to buckling or instability, the possibility of such a failure should be considered in their designs if necessary, adequate bracing must be provided to avoid such failures.
- Trusses: Trusses are composed of straight members connected at their ends by hinged connections to form a stable configuration. Because of their light weight and high strength, are among the most commonly used type of structure.
- Shear structures: These are structures such as reinforced concrete or wooden shear walls, which are used in multistory buildings to reduce lateral movements due to wind loads and earthquake excitations. Shear structures develop mainly in-plane shear with relatively small bending stresses under the action of external loads.
- Bending structures: Bending structures develop mainly bending stresses under the action of external loads. The shear stresses associated with the changes in bending moments may also be significant should be considered in their designs.

High-Rise Buildings

The structural system of a high-rise building is designed to cope with the vertical gravity loads and lateral loads caused by wind or seismic activity. The structural system consists only of the members designed to carry the loads, all other members are referred to as non-structural.

A classification for the structural system of a high-rise was introduced in 1969 by Fazlur Khan and was extended to incorporate interior and exterior structures. The primary lateral load-resisting system defines if a structural system is an interior or exterior one. The following interior structures are possible:

- Hinged frame
- Rigid frame
- Braced frame and Shear-walled frame

- Outrigger structures

The following exterior structures are possible:

- Tube
- Diagrid
- Space Truss
- Exoskeleton
- Superframe

Rigid Frame

A **rigid frame** in structural engineering is the load-resisting skeleton constructed with straight or curved members interconnected by mostly rigid connections which resist movements induced at the joints of members. Its members can take bending moment, shear, and axial loads. The two common assumptions as to the behavior of a building frame are (1) that its beams are free to rotate at their connections or (2) that its members are so connected that the angles they make with each other do not change under load. Frame works with connections of intermediate stiffness will be intermediate between these two extremes. Frame works with connections of intermediate stiffness are commonly called semirigid frames. The AISC specifications recognize three basic frame types: Rigid Frame, Simple Frame, and Partially Restrained Frame.

The AISC Steel Specification Commentary on Section B3 provides guidance for the classification of a connection in terms of its rigidity. The secant stiffness of the connection K_s is taken as an index property of connection stiffness. Specifically,

$Ks = Ms/\theta s$

where

Ms = moment at service loads, kip-in (N-mm)

θs = rotation at service loads, rads

The secant stiffness of the connection is compared to the rotational stiffness of the connected member as follows, in which L and EI are the length and bending rigidity, respectively, of the beam.

If $K_sL/EI \geq 20$, it is acceptable to consider the connection to be fully restrained (in other words, able to maintain the angles between members). If $K_sL/EI \leq 2$, it is acceptable to consider the connection to be simple (in other words, it rotates without developing moment). Connections with stiffnesses between these two limits are partially restrained and the stiffness, strength and ductility of the connection must be considered in the design.

Braced Frame

A braced frame is a structural system which is designed primarily to resist wind and earthquake forces. Members in a braced frame are designed to work in tension and compression, similar to a truss. Braced frames are almost always composed of steel members.

Types of Braced Frame

The John Hancock Center is a braced tube structure.

Most braced frames are concentric. This means that, where members intersect at a node, the centroid of each member passes through the same point.

Concentrically braced frames can further be classified as either **ordinary** or **special**. Ordinary concentric braced frames (OCBFs) do not have extensive requirements regarding members or connections, and are frequently used in areas of low seismic risk. OCBF steel frame buildings originated in Chicago and reinforced concrete frames originated in Germany and France – areas where earthquakes were not an engineering consideration. Accordingly, special concentrically or eccentrically braced frames were later developed with extensive design requirements, and are frequently used in areas of high seismic risk. The purpose of the concentrically- or eccentrically-braced design is to ensure adequate ductility (i.e., to stretch without breaking suddenly).

Outrigger

An **outrigger** is a projecting structure on a boat, with specific meaning depending on types of vessel.

Relief of Borobudur Temple (8th century AD) in Central Java, Indonesia, showing a ship with outrigger

Outrigger on a contemporary Hawaiian sailing canoe

Powered Vessels and Sailboats

An outrigger describes any contraposing float rigging beyond the side of a boat to improve the vessel's stability. If a single outrigger is used it is usually but not always windward.

In an outrigger canoe and in sailboats such as the proa, an outrigger is a thin, long, solid, hull used to stabilise an inherently unstable main hull. The outrigger is positioned rigidly and parallel to the main hull so that the main hull is less likely to capsize. If only one outrigger is used on a vessel, its weight reduces the tendency to capsize in one direction and its buoyancy reduces the tendency in the other direction.

On a keelboat, "outrigger" refers to a variety of structures by which the running rigging (such as a sheet) may be attached outboard (outside the lateral limits) of the boat's hull. The Racing Rules of Sailing generally prohibit such outriggers, though they are explicitly permitted on specific classes, such as the IMOCA Open 60 used in several major offshore races.

Fishing

In fishing from vessels, an outrigger is a pole or series of poles that allow boats to trawl more lines in the water without tangling and simulates a school of fish.

Rowing

Early racing sculls with outriggers in 1851.

In a rowing boat or galley, an outrigger (or just rigger) is a triangular frame that holds the rowlock (into which the oar is slotted) away from the saxboard (gunwale for gig rowing) to optimize leverage. Wooden outriggers appear on the new trireme around the 7th or 6th centuries BC and later on Italian galleys around AD 1300 while Harry

Clasper (1812–1870), a British professional rower, popularised the use of the modern metal version and the top rowing events accepted the physiological and ergonomic advantages so acceded to its use in competitions. **Wing-riggers** are made by some manufacturers of racing shells which are reinforced arcs or form a single projection akin to aircraft wings instead of conventional thin metal triangular structures.

Tube (Structure)

In structural engineering, the tube is the system where in order to resist lateral loads (wind, seismic, etc.) a building is designed to act like a hollow cylinder, cantilevered perpendicular to the ground. This system was introduced by Fazlur Rahman Khan while at Skidmore, Owings and Merrill's (SOM) Chicago office. The first example of the tube's use is the 43-story Khan-designed DeWitt-Chestnut Apartment Building in Chicago, Illinois, completed in 1963.

The system can be constructed using steel, concrete, or composite construction (the discrete use of both steel and concrete). It can be used for office, apartment and mixed-use buildings. Most buildings in excess of 40 stories constructed since the 1960s are of this structural type.

John Hancock Center in Chicago, designed in 1965 and completed in 1969, is an example of the trussed tube structural design

Concept

The tube system concept is based on the idea that a building can be designed to resist lateral loads by designing it as a hollow cantilever perpendicular to the ground. In the simplest incarnation of the tube, the perimeter of the exterior consists of closely spaced columns that are tied together with deep spandrel beams through moment connections. This assembly of columns and beams forms a rigid frame that amounts to a dense and strong structural wall along the exterior of the building.

This exterior framing is designed sufficiently strong to resist all lateral loads on the building, thereby allowing the interior of the building to be simply framed for gravity loads. Interior columns are comparatively few and located at the core. The distance between the exterior and the core frames is spanned with beams or trusses and intentionally left column-free. This maximizes the effectiveness of the perimeter tube by transferring some of the gravity loads within the structure to it and increases its ability to resist overturning due to lateral loads.

By 1963, a new structural system of framed tubes had appeared in skyscraper design and construction. Fazlur Rahman Khan defined the framed tube structure as "a three dimensional space structure composed of three, four, or possibly more frames, braced frames, or shear walls, joined at or near their edges to form a vertical tube-like structural system capable of resisting lateral forces in any direction by cantilevering from the foundation." Closely spaced interconnected exterior columns form the tube. Horizontal loads, like wind for example, are supported by the structure as a whole. About half the exterior surface is available for windows. Framed tubes allow fewer interior columns, and so create more usable floor space. Where larger openings like garage doors are required, the tube frame must be interrupted, with transfer girders used to maintain structural integrity.

The first building to apply the tube-frame construction was the DeWitt-Chestnut apartment building which Khan designed and which was completed in Chicago by 1963. This laid the foundations for the tube structural design of many later skyscrapers, including his own John Hancock Center and Willis Tower, and the construction of the World Trade Center, Petronas Towers, Jin Mao Building, and most other supertall skyscrapers since the 1960s, including the world's current tallest building, the Burj Khalifa.

Variations

From its conception, the tube has been varied to suit different structural requirements:

Framed Tube

The World Trade Center's Twin Towers were some of the first structures to use a framed tube design. The many columns of the framed tube can be seen around the exterior of this horizontal cross section. The towers had a core for services, seen in the center. The design was not tube-in-tube since the core had 47 columns spaced relatively evenly, rather than around the edge of the core.

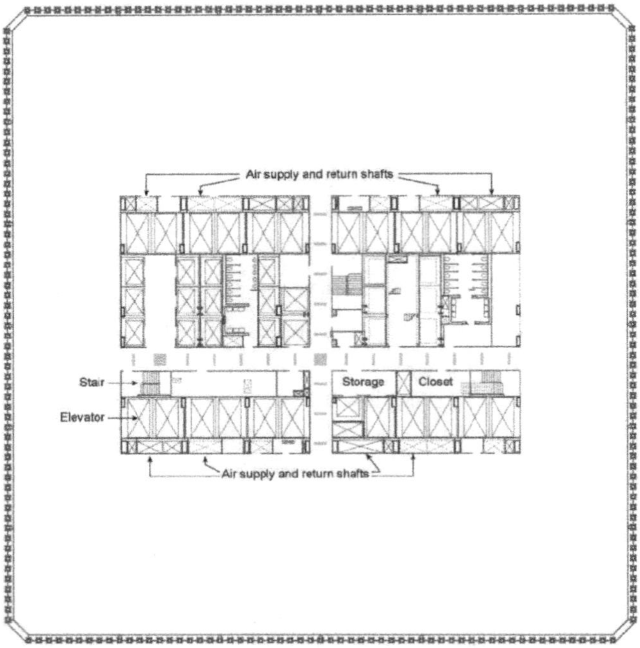

This is the simplest incarnation of the tube. It can appear in a variety of floor plan shapes, including square, rectangular, circular, and freeform. This design was first used in Chicago's DeWitt-Chestnut apartment building, designed by Khan and completed in 1965, but the most notable examples are the Aon Center and the original World Trade Center towers.

Trussed/Braced Tube

Also known as the *braced tube*, the trussed tube is similar to the simple tube but with comparatively fewer and farther-spaced exterior columns. Steel bracings or concrete shear walls are introduced along the exterior walls to compensate for the fewer columns by tying them together. The most notable examples incorporating steel bracing are the John Hancock Center, the Citigroup Center and the Bank of China Tower.

Tube in Tube

Also known as *hull and core*, these structures have a core tube inside the structure, holding the elevator and other services, and another tube around the exterior. The majority of the gravity and lateral loads are normally taken by the outer tube because of its greater strength. 780 Third Avenue, a 50-story concrete frame office building in Manhattan, uses concrete shear walls for bracing and an off-center core to allow column-free interiors.

Bundled Tube

Breakdown of the bundled tube structure of the Willis Tower with simplified floor plans.

Instead of one tube, a building consists of several tubes tied together to resist the lateral forces. Such buildings have interior columns along the perimeters of the tubes when they fall within the building envelope. Notable examples include Willis Tower, One Magnificent Mile, and the Newport Tower.

Willis Tower, completed in 1973, introduced the bundled tube structural design and was the world's tallest building until 1998

The bundle tube design was not only highly efficient in economic terms, but it was also "innovative in its potential for versatile formulation of architectural space. Efficient towers no longer had to be box-like; the tube-units could take on various shapes and could be bundled together in different sorts of groupings." The bundled tube structure meant that "buildings no longer need be boxlike in appearance: they could become sculpture."

Hybrid System

A varied category of structures where the basic concept of tube is used, but has been supplemented by other structural support.This method is used where the slenderness of the building is such that a single system cannot provide adequate strength or stiffness.

Diagram

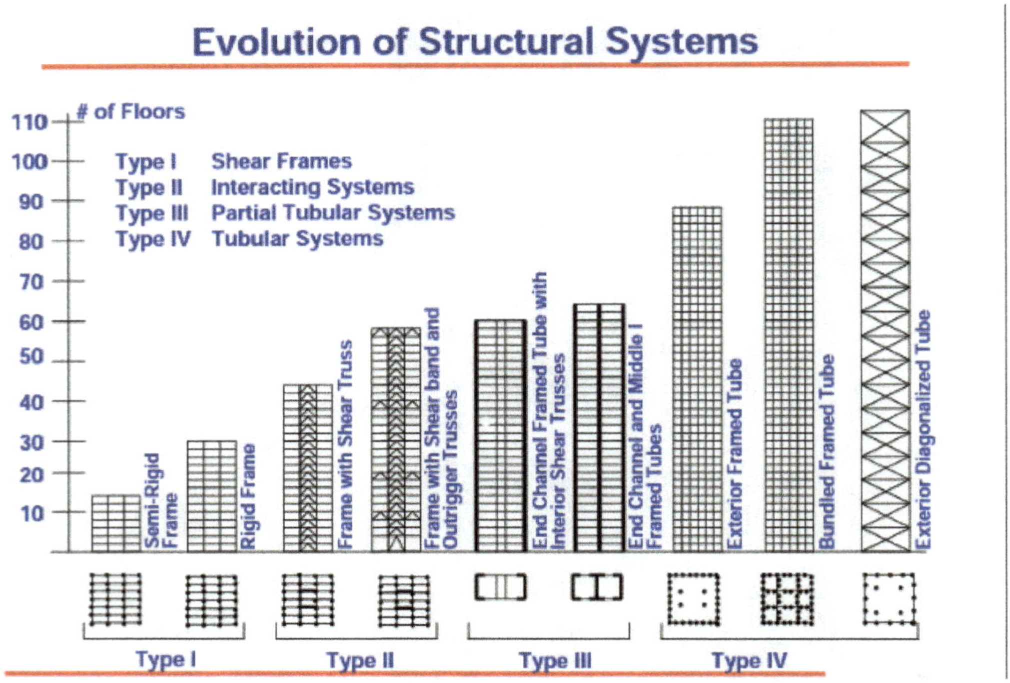

Diagrid

The **diagrid** (a portmanteau of **diagonal grid**) is a framework of diagonally intersecting metal, concrete or wooden beams that is used in the construction of buildings and roofs. It requires less structural steel than a conventional steel frame. Hearst Tower in New York City, designed by Sir Norman Foster, uses 21 percent less steel than a standard design. The diagrid obviates the need for columns and can be used to make large column-free expanses of roofing. Another iconic building designed by Sir Norman Foster, 30 St Mary Axe, known as "The Gherkin", also uses the diagrid system.

Base of 30 St Mary Axe, London, UK

The world's first diagrid hyperboloid structure in Polibino, Russia

MyZeil, Frankfurt, Germany

Buildings Utilizing Diagrid

- Shukhov Tower in Polibino, Polibino, Russia (1896)
- Shukhov Rotunda at the All-Russia exhibition, Nizhny Novgorod, Russia (1896)
- The world's first double curvature steel diagrid roof, Vyksa, Russia, (1897)
- Altair, Colombo, Sri Lanka
- Shukhov Tower, Moscow, Russia
- Hearst Tower, New York, USA
- 30 St Mary Axe, London, England
- 1 The Avenue, Manchester, England
- CCTV Headquarters, Beijing, China
- The Bow, Calgary, Canada
- Seattle Central Library, Seattle, USA
- Capital Gate, Abu Dhabi, United Arab Emirates
- Aldar headquarters, Abu Dhabi, United Arab Emirates
- Guangzhou International Finance Center, Guangzhou, China
- Queen Elizabeth II Great Court at the British Museum, London, England
- Nagoya Dome, Nagoya, Japan
- Korkeasaari Lookout Tower, Helsinki, Finland
- Milan Trade Fair, Milan, Italy
- Westhafen Tower, Frankfurt, Germany
- MyZeil, Frankfurt, Germany
- The Crystal, Copenhagen, Denmark

- Botanical Garden Shanghai, Shanghai, China
- Paunsdorf Center, Leipzig, Germany
- Coastlands Aquatic Centre, Kapiti, New Zealand
- South Australian Health and Medical Research Institute, Adelaide, Australia
- Departures concourse at King's Cross railway station, London, England
- United Steelworkers Building, Pittsburgh, USA
- Tornado Tower, Doha, Qatar
- One Shelley Street, Sydney, Australia
- Transport interchange at the Manukau Institute of Technology, Otara, New Zealand
- The WWF Living Planet Centre, Woking, England
- Newfoundland Quay, London, England

REFERENCES

1. "What is a structural engineer". RMG Engineers. 2015-11-30. Retrieved 2015-11-30.
2. IABSE "Organisation", iabse website
3. Victor E. Saouma. "Lecture notes in Structural Engineering" (PDF). University of Colorado. Retrieved 2007-11-02.
4. Fonte, Gerard C. A. Building the Great Pyramid in a Year : An Engineer's Report (Report). Algora Publishing: New York. p. 34.
5. unknown. "Some Useful Numbers on the Engineering Properties of Materials (Geologic and Otherwise)" (PDF). Stanford University. Retrieved 2013-12-05.
6. "ETABS receives "Top Seismic Product of the 20th Century" Award" (PDF). Press Release. Structure Magazine. 2006. Retrieved April 20, 2012.
7. http://www.ramsay-maunder.co.uk/downloads/precast_roof_slabs.pdf
8. http://www.ramsay-maunder.co.uk/downloads/l_shaped_landing.pdf

CHAPTER 4

Transportation Engineering

Chapter Contents

Railway, bus, highway, and airport systems are intricate infrastructures that require considerable planning and development. A person who designs these systems is known as a transportation engineer. Transportation engineers may work for a variety of sectors, including consulting firms, private businesses, universities, and government agencies.

INTRODUCTION

Transportation engineering or transport engineering is the application of technology and scientific principles to the planning, functional design, operation and management of facilities for any mode of transportation in order to provide for the safe, efficient, rapid, comfortable, convenient, economical, and environmentally compatible movement of people and goods (transport). It is a sub-discipline of civil engineering. The importance of transportation engineering within the civil engineering profession can be judged by the number of divisions in ASCE (American Society of Civil Engineers) that are directly related to transportation. There are six such divisions (Aerospace; Air Transportation; Highway; Pipeline; Waterway, Port, Coastal and Ocean; and Urban Transportation) representing one-third of the total 18 technical divisions within the ASCE (1987).

The planning aspects of transportation engineering relate to elements of urban planning, and involve technical forecasting decisions and political factors. Technical forecasting of passenger travel usually involves an urban transportation planning model, requiring the estimation of trip generation (how many trips for what purpose), trip distribution (destination choice, where is the traveler going), mode choice (what mode is being taken), and route assignment (which streets or routes are being used). More sophisticated forecasting can include other aspects of traveler decisions, including auto ownership, trip chaining (the decision to link individual trips together in a tour) and the choice of residential or business location (known as land use forecasting). Passenger trips are the focus of transportation engineering because they often represent the peak of demand on any transportation system.

A review of descriptions of the scope of various committees indicates that while facility planning and design continue to be the core of the transportation engineering field, such areas as operations planning, logistics, network analysis, financing, and policy analysis are also important to civil engineers, particularly to those working in highway and urban transportation. The National Council of Examiners for Engineering and Surveying (NCEES) list online the safety protocols, geometric design requirements, and signal timing.

Transportation engineering, as practiced by civil engineers, primarily involves planning, design, construction, maintenance, and operation of transportation facilities. The facilities support air, highway, railroad, pipeline, water, and even space transportation. The design aspects of transportation engineering include the sizing of transportation facilities (how many lanes or how much capacity the facility has), determining the materials and thickness used in pavement designing the geometry (vertical and horizontal alignment) of the roadway (or track).

Before any planning occurs the Engineer must take what is known as an inventory of the area or if it is appropriate, the previous system in place. This inventory or database must include information on (1)population, (2)land use, (3)economic activity, (4)transportation facilities and services, (5)travel patterns and volumes, (6)laws and ordinances, (7)regional financial resources, (8)community values and expectations. These inventories help the engineer create business models to complete accurate forecasts of the future conditions of the systemReview.

Operations and management involve traffic engineering, so that vehicles move smoothly on the road or track. Older techniques include signs, signals, markings, and tolling. Newer technologies involve intelligent transportation systems, including advanced traveler information systems (such as variable message signs), advanced traffic control systems (such as ramp meters), and vehicle infrastructure integration. Human factors are an aspect of transportation engineering, particularly concerning driver-vehicle interface and user interface of road signs, signals, and markings.

The engineering of this roundabout in Bristol, England, attempts to make traffic flow free-moving

Highway Engineering

Engineers in this specialization:

- Handle the planning, design, construction, and operation of highways, roads, and other vehicular facilities as well as their related bicycle and pedestrian realms.
- Estimate the transportation needs of the public and then secure the funding for the project.
- Analyze locations of high traffic volumes and high collisions for safety and capacity.
- Use civil engineering principles to improve the transportation system.
- Utilizes the three design controls which are the drivers, the vehicles, and the roadways themselves.

HIGHWAY ENGINEERING

Highway engineering is an engineering discipline branching from civil engineering that involves the planning, design, construction, operation, and maintenance of roads, bridges, and tunnels to ensure safe and effective transportation of people and goods. Highway engineering became prominent towards the latter half of the 20th Century after World War 2. Standards of highway engineering are continuously being improved. Highway engineers must take into account future traffic flows, design of highway intersections/ interchanges, geometric alignment and design, highway pavement materials and design, structural design of pavement thickness, and pavement maintenance.

German Autobahn, 1936–1939

The beginning of road construction could be dated to the time of the Romans. With the advancement of technology from carriages pulled by two horses to vehicles with power equivalent to 100 horses, road development had to follow suit. The construction of modern highways did not begin until the late 19th to early 20th century.

The first research dedicated to highway engineering was initiated in the United Kingdom with the introduction of the Transport Research Laboratory (TRL), in 1930. In the USA, highway engineering became an important discipline with the passing of the Federal-Aid Highway Act of 1944, which aimed to connect 90% of cities with a population of 50,000 or more. With constant stress from vehicles which grew larger as time passed, improvements to pavement were needed. With technology out of date, in 1958 the construction of the first motorway in Great Britain (the Preston bypass) played a major role in the development of new pavement technology.

Design policies standards used in the United States are typically based on publications of the American Association of State Highway and Transportation Officials as well as research promulgated by the Transportation Research Board, the Institute of Transportation Engineers, the Federal Highway Administration, and the Department of Transportation.

Planning and Development

Highway planning involves the estimation of current and future traffic volumes on a road network. Highway engineers strive to predict and analyze all possible civil impacts of highway systems. Some considerations are the adverse effects on the environment, such as noise pollution, air pollution, water pollution, and other ecological impacts.

Financing

Developed countries are constantly faced with high maintenance cost of aging transportation highways. The growth of the motor vehicle industry and accompanying economic growth has generated a demand for safer, better performing, less congested highways. The growth of commerce, educational institutions, housing, and defense have largely drawn from government budgets in the past, making the financing of public highways a challenge.

The multipurpose characteristics of highways, economic environment, and the advances in highway pricing technology are constantly changing. Therefore, the approaches to highway financing, management, and maintenance are constantly changing as well.

Environmental Impact Assessment

The economic growth of a community is dependent upon highway development to enhance mobility. However, improperly planned, designed, constructed, and maintained highways can disrupt the social and economic characteristics of any size community. Common adverse impacts to highway development include damage of habitat and bio-diversity, creation of air and water pollution, noise/vibration generation, damage of natural landscape, and the destruction of a community's social and cultural structure. Highway infrastructure must be constructed and maintained to high qualities and standards.

There are three key steps for integrating environmental considerations into the planning, scheduling, construction, and maintenance of highways. This process is known as an Environmental Impact Assessment, or EIA, as it systematically deals with the following elements:

- Identification of the full range of possible impacts on the natural and socio-economic environment
- Evaluation and quantification of these impacts
- Formulation of measures to avoid, mitigate, and compensate for the anticipated impacts.

Highway Safety

Highway systems generate the highest price in human injury and death, as nearly 50 million persons are injured in traffic accidents every year, not including the 1.2 million deaths. Road traffic injury is the single leading cause of unintentional death in the first five decades of human life.

Management of safety is a systematic process that strives to reduce the occurrence and severity of traffic accidents. The man/machine interaction with road traffic systems is unstable and poses a challenge to highway safety management. The key for increasing the safety of highway systems is to design, build, and maintain them to be far more tolerant of the average range of this man/machine interaction with highways. Technological advancements in highway engineering have improved the design, construction, and maintenance methods used over the years. These advancements have allowed for newer highway safety innovations.

By ensuring that all situations and opportunities are identified, considered, and implemented as appropriate, they can be evaluated in every phase of highway planning, design, construction, maintenance, and operation to increase the safety of our highway systems.

Design

The most appropriate location, alignment, and shape of a highway are selected during the design stage. Highway design involves the consideration of three major factors (human, vehicular, and roadway) and how these factors interact to provide a safe highway. Human factors include reaction time for braking and steering, visual acuity for traffic signs and signals, and car-following behaviour. Vehicle considerations include vehicle size and dynamics that are essential for determining lane width and maximum slopes, and for the selection of design vehicles. Highway engineers design road geometry to ensure stability of vehicles when negotiating curves and grades and to provide adequate sight distances for undertaking passing maneuvers along curves on two-lane, two-way roads.

Geometric Design

Highway and transportation engineers must meet many safety, service, and performance standards when designing highways for certain site topography. Highway geometric design primarily refers to the visible elements of the highways. Highway engineers who design the geometry of highways must also consider environmental and social effects of the design on the surrounding infrastructure.

There are certain considerations that must be properly addressed in the design process to successfully fit a highway to a site's topography and maintain its safety. Some of these design considerations include:

- Design speed
- Design traffic volume
- Number of lanes
- Level of Service (LOS)
- Sight Distance
- Alignment, super-elevation, and grades
- Cross section
- Lane width
- Horizontal and vertical clearance

The operational performance of a highway can be seen through drivers' reactions to the design considerations and their interaction.

Materials

The materials used for roadway construction have progressed with time, dating back to the early days of the Roman Empire. Advancements in methods with which these materials are characterized and applied to pavement structural design has accompanied this advancement in materials.

There are two major types of pavement surfaces - Portland cement concrete (PCC) and hot-mix asphalt (HMA). Underneath this wearing course are material layers that give structural support for the pavement system. These underlying surfaces may include either the aggregate base and subbase layers, or treated base and subbase layers, and additionally the underlying natural or treated subgrade. These treated layers may be cement-treated, asphalt-treated, or lime-treated for additional support.

Flexible Pavement Design

A flexible, or asphalt, or Tarmac pavement typically consists of three or four layers. For a four layer flexible pavement, there is a surface course, base course, and subbase course constructed over a compacted, natural soil subgrade. When building a three layer flexible pavement, the subbase layer is not used and the base course is placed directly on the natural subgrade.

A flexible pavement's surface layer is constructed of hot-mix asphalt (HMA). Unstabilized aggregates are typically used for the base course; however, the base course could also be stabilized with asphalt, Foamed Bitumen,<Roadstone Recycling> Portland cement, or another stabilizing agent. The subbase is generally constructed from local aggregate material, while the top of the subgrade is often stabilized with cement or lime.

With flexible pavement, the highest stress occurs at the surface and the stress decreases as the depth of the pavement increases. Therefore, the highest quality material needs to be used for the surface, while lower quality materials can be used as the depth of the pavement increases. The term "flexible" is used because of the asphalts ability to bend and deform slightly, then return to its original position as each traffic load is applied and removed. It is possible for these small deformations to become permanent, which can lead to rutting in the wheel path over an extended time.

The service life of a flexible pavement is typically designed in the range of 15 to 20 years. Required thicknesses of each layer of a flexible pavement vary widely depending on the materials used, magnitude, number of repetitions of traffic loads, environmental conditions, and the desired service life of the pavement. Factors such as these are taken into consideration during the design process so that the pavement will last for the designed life without excessive distresses.

Rigid Pavement Design

Rigid pavements are generally used in constructing airports and major highways, such as those in the interstate highway system. In addition, they commonly serve as heavy-duty industrial floor slabs, port and harbor yard pavements, and heavy-vehicle park or terminal pavements. Like flexible pavements, rigid highway pavements are designed as all-weather, long-lasting structures to serve modern day high-speed traffic. Offering high quality riding surfaces for safe vehicular travel, they function as structural layers to distribute vehicular wheel loads in such a manner that the induced stresses transmitted to the subgrade soil are of acceptable magnitudes.

Portland cement concrete (PCC) is the most common material used in the construction of rigid pavement slabs. The reason for its popularity is due to its availability and the economy. Rigid pavements must be designed to endure frequently repeated traffic loadings. The typical designed service life of a rigid pavement is between 30 and 40 years, lasting about twice as long as a flexible pavement.

One major design consideration of rigid pavements is reducing fatigue failure due to the repeated stresses of traffic. Fatigue failure is common among major roads because a typical highway will experience millions of wheel passes throughout its service life. In addition to design criteria such as traffic loadings, tensile stresses due to thermal energy must also be taken into consideration. As pavement design has progressed, many highway engineers have noted that thermally induced stresses in rigid pavements can be just as intense as those imposed by wheel loadings. Due to the relatively low tensile strength of concrete, thermal stresses are extremely important to the design considerations of rigid pavements.

Rigid pavements are generally constructed in three layers - a prepared subgrade, base or subbase, and a concrete slab. The concrete slab is constructed according to a designed choice of plan dimensions for the slab panels, directly influencing the intensity of thermal stresses occurring within the pavement. In addition to the slab panels, temperature reinforcements must be designed to control cracking behavior in the slab. Joint spacing is determined by the slab panel dimensions.

Three main types of concrete pavements commonly used are Jointed plain concrete pavement (JPCP), jointed reinforced concrete pavement (JRCP), and continuously reinforced concrete pavements (CRCP). JPCP's are constructed with contraction joints which direct the natural cracking of the pavement. These pavements do not use any reinforcing steel. JRCP's are constructed with both contraction joints and reinforcing steel to control the cracking of the pavement. High temperatures and moisture stresses within the pavement creates cracking, which the reinforcing steel holds tightly together. At transverse joints, dowel bars are typically placed to assist with transferring the load of the vehicle across the cracking. CRCP's solely rely on continuous reinforcing steel to hold the pavement's natural transverse cracks together. Prestressed concrete pavements have also been used in the construction of highways; however, they are not as common as the other three. Prestressed pavements allow for a thinner slab thickness by partly or wholly neutralizing thermally induced stresses or loadings.

Flexible Pavement Overlay Design

Over the service life of a flexible pavement, accumulated traffic loads may cause excessive rutting or cracking, inadequate ride quality, or an inadequate skid resistance. These problems can be avoided by adequately maintaining the pavement, but the solution usually has excessive maintenance costs, or the pavement may have an inadequate structural capacity for the projected traffic loads.

Throughout a highway's life, its level of serviceability is closely monitored and maintained. One common method used to maintain a highway's level of serviceability is to place an overlay on the pavement's surface.

There are three general types of overlay used on flexible pavements: asphalt-concrete overlay, Portland cement concrete overlay, and ultra-thin Portland cement concrete overlay. The concrete layer in a conventional PCC overlay is placed unbonded on top of the flexible surface. The typical thickness of an ultra-thin PCC overlay is 4 inches (10 cm) or less.

There are three main categories of flexible pavement overlay design procedures:

- Component Analysis Design
- Deflection-Based Design

Rigid Pavement Overlay Design

Near the end of a rigid pavement's service life, a decision must be made to either fully reconstruct the worn pavement, or construct an overlay layer. Considering an overlay can be constructed on a rigid pavement that has not reached the end of its service life, it is often more economically attractive to apply overlay layers more frequently. The required overlay thickness for a structurally sound rigid pavement is much smaller than for one that has reached the end of its service life. Rigid and flexible overlays are both used for rehabilitation of rigid pavements such as JPCP, JRCP, and CRCP.

There are three subcategories of rigid pavement overlays that are organized depending on the bonding condition at the pavement overlay and existing slab interface.

- Bonded overlays
- Unbonded overlays
- Partially bonded overlays

Drainage System Design

Designing for proper drainage of highway systems is crucial to their success. Regardless of how well other aspects of a road are designed and constructed, adequate drainage is mandatory for a road to survive its entire service life. Excess water in the highway structure can inevitably lead to premature failure, even if the failure is not catastrophic.

Each highway drainage system is site-specific and can be very complex. Depending on the geography of the region, many methods for proper drainage may not be applicable. The highway engineer must determine which situations a particular design process should be applied, usually a combination of several appropriate methods and materials to direct water away from the structure.

Erosion control is a crucial component in the design of highway drainage systems. Surface drainage must be allowed for precipitation to drain away from the structure. Highways must be designed with a slope or crown so that runoff water will be directed to the shoulder of the road, into a ditch, and away from the site. Designing a drainage system requires the prediction of runoff and infiltration, open channel analysis, and culvert design for directing surface water to an appropriate location.

Construction, Maintenance, and Management

Highway Construction

Highway construction is generally preceded by detailed surveys and subgrade preparation. The methods and technology for constructing highways has evolved over time and become increasingly sophisticated. This advancement in technology has raised the level of skill sets required to manage highway construction projects. This skill varies from project to project, depending on factors such as the project's complexity and nature, the contrasts between new construction and reconstruction, and differences between urban region and rural region projects.

There are a number of elements of highway construction which can be broken up into technical and commercial elements of the system. Some examples of each are listed below:

- Technical Elements
  - Materials
  - Material quality
  - Installation techniques
  - Traffic
- Commercial Elements
  - Contract understanding
  - Environmental aspects
  - Political aspects
  - Legal aspects
  - Public concerns

Typically, construction begins at the lowest elevation of the site, regardless of the project type, and moves upward. By reviewing the geotechnical specifications of the project, information is given about:

- Existing ground conditions
- Required equipment for excavation, grading, and material transportation to and from the site
- Properties of materials to be excavated
- Dewatering requirements necessary for below-grade work
- Shoring requirements for excavation protection
- Water quantities for compaction and dust control

Subbase Course Construction

A subbase course is a layer designed of carefully selected materials that is located between the subgrade and base course of the pavement. The subbase thickness is generally in the range of 4 to 16 inches, and it is designed to withstand the required structural capacity of the pavement section.

Common materials used for a highway subbase include gravel, crushed stone, or subgrade soil that is stabilized with cement, fly ash, or lime. Permeable subbase courses are becoming more prevalent because of their ability to drain infiltrating water from the surface. They also prevent subsurface water from reaching the pavement surface.

When local material costs are excessively expensively or the material requirements to increase the structural bearing of the sub-base are not readily available, highway engineers can increase the bearing capacity of the underlying soil by mixing in Portland cement, foamed asphalt, or with emerging technologies such as the cross-linking styrene acrylic polymer that increases the California Bearing Ratio of in-situ materials by a factor 4 – 6.

Base Course Construction

The base course is the region of the pavement section that is located directly under the surface course. If there is a subbase course, the base course is constructed directly about this layer. Otherwise, it is built directly on top of the subgrade. Typical base course thickness ranges from 4 to 6 inches and is governed by underlying layer properties.

Heavy loads are continuously applied to pavement surfaces, and the base layer absorbs the majority of these stresses. Generally, the base course is constructed with an untreated crushed aggregate such as crushed stone, slag, or gravel. The base course material will have stability under the construction traffic and good drainage characteristics.

The base course materials are often treated with cement, bitumen, calcium chloride, sodium chloride, fly ash, or lime. These treatments provide improved support for heavy loads, frost susceptibility, and serves as a moisture barrier between the base and surface layers.

Surface Course Construction

There are two most commonly used types of pavement surfaces used in highway construction: hot-mix asphalt and Portland cement concrete. These pavement surface courses provide a smooth and safe riding surface, while simultaneously transferring the heavy traffic loads through the various base courses and into the underlying subgrade soils.

Hot-Mix Asphalt (HMA) Layers

Hot-mix asphalt surface courses are referred to as flexible pavements. The Superpave System was developed in the late 1980s and has offered changes to the design approach, mix design, specifications, and quality testing of materials.

The construction of an effective, long-lasting asphalt pavement requires an experienced construction crew, committed to their work quality and equipment control.

Construction issues:

- Asphalt mix segregation
- Laydown

- Compaction
- Joints

A prime coat is a low viscosity asphalt that is applied to the base course prior to laying the HMA surface course. This coat bonds loose material, creating a cohesive layer between the base course and asphalt surface.

A tack coat is a low viscosity asphalt emulsion that is used to create a bond between an existing pavement surface and new asphalt overlay. Tack coats are typically applied on adjacent pavements (curbs) to assist the bonding of the HMA and concrete.

Portland Cement Concrete (PCC)

Portland cement concrete surface courses are referred to as rigid pavements, or concrete pavements. There are three general classifications of concrete pavements - jointed plain, jointed reinforced, and continuously reinforced.

Traffic loadings are transferred between sections when larger aggregates in the PCC mix inter-lock together, or through load transfer devices in the transverse joints of the surface. Dowel bars are used as load-transferring devices to efficiently transfer loads across transverse joints while maintaining the joint's horizontal and vertical alignment. Tie-bars are deformed steel bars that are placed along longitudinal joints to hold adjacent pavement sections in place.

Highway Maintenance

The overall purpose of highway maintenance is to fix defects and preserve the pavement's structure and serviceability. Defects must be defined, understood, and recorded in order to select an appropriate maintenance plan. Defects differ between flexible and rigid pavements.

There are four main objectives of highway maintenance:

- repair of functional pavement defects
- extend the functional and structural service life of the pavement
- maintain road safety and signage
- keep road reserve in acceptable condition

Through routine maintenance practices, highway systems and all of their components can be maintained to their original, as-built condition.

Project Management

Project management involves the organization and structuring of project activities from inception to completion. Activities could be the construction of infrastructure such as highways and bridges or major and minor maintenance activities related to constructing such infrastructure. The entire project and involved activities must be handled in a professional manner and completed within deadlines and budget. In addition, minimizing social and environmental impacts is essential to successful project management.

RAILWAY ENGINEERING

Railway engineering is a multi-faceted engineering discipline dealing with the design, construction and operation of all types of railway systems. It encompasses a wide range of engineering disciplines, including civil engineering, computer engineering, electrical engineering, mechanical engineering, industrial engineering and production engineering. A great many other engineering sub-disciplines are also called upon.

With the advent of the railways in the early nineteenth century, a need arose for a specialized group of engineers capable of dealing with the unique problems associated with railway engineering. As the railways expanded and became a major economic force, a great many engineers became involved in the field, probably the most notable in Britain being Richard Trevithick, George Stephenson and Isambard Kingdom Brunel. Today, railway systems engineering continues to be a vibrant field of engineering.

Subfields

- Mechanical engineering
- Command, control & railway signalling
  - Office systems design
  - Data center design
  - SCADA
  - Network design
- Electrical engineering
  - Energy electrification
    - Third rail
      - Fourth rail
    - Overhead contact system
- Civil engineering
  - Permanent way engineering
- Light rail systems
- On-track plant
- Rail systems integration
- Train control systems
  - Cab signalling
- Railway vehicle engineering
- Rolling resistance
  - Curve resistance

- Wheel–rail interface
  - Hunting oscillation
- Railway systems engineering
  - Railway signalling
  - Fare collection
  - CCTV
  - Public address
  - Intrusion detection
  - Access control
- Systems integration

Professional Organizations

In the UK: The Railway Division of the Institution of Mechanical Engineers (IMechE).

In the US The American Railway Engineering and Maintenance-of-Way Association (AREMA)

Worldwide The Institute of Railway Signal Engineers (IRSE)

Railroad Engineering

Railway engineers handle the design, construction, and operation of railroads and mass transit systems that use a fixed guideway (such as light rail or even monorails). Typical tasks would include determining horizontal and vertical alignment design, station location and design, and construction cost estimating. Railroad engineers can also move into the specialized field of train dispatching which focuses on train movement control.

Railway engineers also work to build a cleaner and safer transportation network by reinvesting and revitalizing the rail system to meet future demands. In the United States, railway engineers work with elected officials in Washington, D.C. on rail transportation issues to make sure that the rail system meets the country's transportation needs.

Port and Harbor Engineering

Port and harbor engineers handle the design, construction, and operation of ports, harbors, canals, and other maritime facilities. This is not to be confused with marine engineering.

Airport Engineering

Airport engineers design and construct airports. Airport engineers must account for the impacts and demands of aircraft in their design of airport facilities. These engineers must use the analysis of predominant wind direction to determine runway orientation, determine the size of runway border and safety areas, different wing tip to wing tip clearances for all gates and must designate the clear zones in the entire port.

Professional Societies

- Young Professionals in Transportation (YPT)
- Institute of Transportation Engineers (ITE)
- American Society for Highway Engineering (ASHE)
- American Society of Civil Engineers (ASCE)
- The American Association of State Highway and Transportation Officials (AASHTO)
- Eastern Asia Society for Transportation Studies (EASTS)

REFERENCES

1. Rogers, Martin (2002). Highway engineering. Oxford, UK: Blackwell Science. ISBN 978-0-632-05993-5.
2. O'Flaherty, edited by C.A. (2002). Highways the location, design, construction and maintenance of road pavements (4th ed.). Oxford: Butterworth-Heinemann. ISBN 978-0-7506-5090-8.
3. "Highway engineering." McGraw-Hill Concise Encyclopedia of Science and Technology. New York: McGraw-Hill, 2006. Credo Reference. Web. 13 February 2013.
4. Chin, Antony T.H. "Financing Highways." The Handbook of Highway Engineering. Ed. T.W. Fwa. CRC Press, 2005.
5. Estache, A., Romero, M., and Strong, J. 2000. The Long and Winding Path to Private Financing and Regulation of Toll Roads. The World Bank, Washington, DC, 49 pp.
6. Aziz, M.A. "Environmental Impact Assessment of Highway Development." The Handbook of Highway Engineering. Ed. T.W. Fwa. CRC Press, 2005.
7. World Health Organization, 2004. World Report on Road Traffic Injury Prevention. World Health Organization, Geneva.
8. Johnston, Ian. "Highway Safety." The Handbook of Highway Engineering. Ed. T.W. Fwa. CRC Press, 2005.
9. Cheu, R.L. "Highway Geometric Design." The Handbook of Highway Engineering. Ed. T.W. Fwa. CRC Press, 2005.

10. Tam, Weng On. "Highway Materials." The Handbook of Highway Engineering. Ed. T.W. Fwa. CRC Press, 2005.
11. Mamlouk, Michael S. "Design of Flexible Pavements." The Handbook of Highway Engineering. Ed. T.W. Fwa. CRC Press, 2005.
12. Fwa, T.F. and Wei, Liu. "Design of Rigid Pavements." The Handbook of Highway Engineering. Ed. T.W. Fwa. CRC Press, 2005.
13. Tia, Mang. "Overlay Design for Flexible Pavements." The Handbook of Highway Engineering. Ed. T.W. Fwa. CRC Press, 2005.
14. Fwa, T.F. "Overlay Design for Rigid Pavements." The Handbook of Highway Engineering. Ed. T.W. Fwa. CRC Press, 2005.
15. Ksaibati, Khaled and Kolkman, Laycee L. "Highway Drainage Systems and Design." The Handbook of Highway Engineering. Ed. T.W. Fwa. CRC Press, 2005.
16. Gunalan, K.N. "Highway Construction." The Handbook of Highway Engineering. Ed. T.W. Fwa. CRC Press, 2005.
17. "AggreBind - Soil Stabilized Roads and Highways".
18. Van Wijk, Ian. "Highway Maintenance." The Handbook of Highway Engineering. Ed. T.W. Fwa. CRC Press, 2005.
19. Walker, Derek and Kumar, Arun. "Project Management in Highway Construction." The Handbook of Highway Engineering. Ed. T.W. Fwa. CRC Press, 2005.
20. "ITE – The Transportation Profession". ITE. Retrieved 2010-06-27.
21. "ASCE – About Civil Engineering". ASCE. Retrieved 2010-06-28.
22. "Association of American Railroads". AAR. Retrieved 2011-06-30.
23. Young Professionals in Transportation. yptransportation.org. Retrieved on 2012-2-13.
24. Institute of Transportation Engineers. ITE. Retrieved on 2011-10-12.
25. American Society for Highway Engineers. highwayengineers.org. Retrieved on 2012-3-4.
26. American Society of Civil Engineers. Asce.org (2010-07-19). Retrieved on 2011-10-12.
27. "American Association for State Highway and Transportation Officials". Retrieved 17 May 2011.

Road Surface Engineering

Chapter Contents

A paved road surface, commonly constructed of bituminous material, should provide a durable, even running surface with adequate skid resistance which is able to withstand the effects of weather. When the road surface condition becomes poor or defective, rehabilitation or resurfacing work is needed to restore the pavement surface to the required level of service.

INTRODUCTION

Road surface or pavement is the durable surface material laid down on an area intended to sustain vehicular or foot traffic, such as a road or walkway. In the past, gravel road surfaces, cobblestone and granite setts were extensively used, but these surfaces have mostly been replaced by asphalt or concrete laid on a compacted base course. Road surfaces are frequently marked to guide traffic. Today, permeable paving methods are beginning to be used for low-impact roadways and walkways.

A road being resurfaced

ASPHALT

Closeup of asphalt on a driveway

Asphalt (specifically, asphalt concrete), sometimes called flexible pavement due to the nature in which it distributes loads, has been widely used since the 1920s. The viscous nature of the bitumen binder allows asphalt concrete to sustain significant plastic deformation, although fatigue from repeated loading over time is the most common failure mechanism. Most asphalt surfaces are laid on a gravel base, which is generally at least as thick as the asphalt layer, although some 'full depth' asphalt surfaces are laid directly on the native subgrade. In areas with very soft or expansive subgrades such as clay or peat, thick gravel bases or stabilization of the subgrade with Portland cement or lime may be required. Polypropylene and polyester geosynthetics have also been used for this purpose and in some northern countries, a layer of polystyrene boards have been used to delay and minimize frost penetration into the subgrade.

Depending on the temperature at which it is applied, asphalt is categorized as hot mix, warm mix, or cold mix. Hot mix asphalt is applied at temperatures over 300 °F (150 °C) with a free floating screed. Warm mix asphalt is applied at temperatures of 200–250 °F (95–120 °C), resulting in reduced energy usage and emissions of volatile organic compounds. Cold mix asphalt is often used on lower volume rural roads, where hot mix asphalt would cool too much on the long trip from the asphalt plant to the construction site.

An asphalt concrete surface will generally be constructed for high-volume primary highways having an average annual daily traffic load greater than 1200 vehicles per day. Advantages of asphalt roadways include relatively low noise, relatively low cost compared with other paving methods, and perceived ease of repair. Disadvantages include less durability than other paving methods, less tensile strength than concrete, the tendency to become slick and soft in hot weather and a certain amount of hydrocarbon pollution to soil and groundwater or waterways.

In the mid-1960s, rubberized asphalt was used for the first time, mixing crumb rubber from used tires with asphalt. While a potential use for tires that would otherwise fill landfills and present a fire hazard, rubberized asphalt has shown greater incidence of wear in freeze-thaw cycles in temperate zones due to non-homogeneous expansion and contraction with non-rubber components. Also, application of rubberized asphalt is more temperature-sensitive, and in many locations can only be applied at certain times of the year.

Study results of the long-term acoustic benefits of rubberized asphalt are inconclusive. Initial application of rubberized asphalt may provide 3–5 decibels (dB) reduction in tire-pavement source noise emissions; however, this translates to only 1–3 decibels (dB) in total traffic noise level reduction (due to the other components of traffic noise). Compared to traditional passive attenuating measures (e.g., noise walls and earth berms), rubberized asphalt provides shorter-lasting and lesser acoustic benefits at typically much greater expense.

Concrete

Concrete roadway in San Jose, California

Concrete surfaces (specifically, Portland cement concrete) are created using a concrete mix of Portland cement, coarse aggregate, sand and water. In virtually all modern mixes there will also be various admixtures added to increase workability, reduce the required amount of water, mitigate harmful chemical reactions and for other beneficial purposes. In many cases there will also be Portland cement substitutes added, such as fly ash. This can reduce the cost of the concrete and improve its physical properties. The material is applied in a freshly mixed slurry, and worked mechanically to compact the interior and force some of the cement slurry to the surface to produce a smoother, denser surface free from honeycombing. The water allows the mix to combine molecularly in a chemical reaction called hydration.

Concrete surfaces have been refined into three common types: jointed plain (JPCP), jointed reinforced (JRCP) and continuously reinforced (CRCP). The one item that distinguishes each type is the jointing system used to control crack development.

- Jointed plain concrete pavements contain enough joints to control the location of all the expected shrinkage cracks. The concrete cracks at the joints and not elsewhere in the slabs. Jointed plain pavements do not contain any steel reinforcement. However, there may be smooth steel bars at transverse joints and deformed steel bars at longitudinal joints. The spacing between transverse joints is typically about 15 feet (4.6 m) for slabs 7 to 12 inches (180 to 300 mm) thick. Today, a majority of US state agencies build jointed plain pavements.
- Jointed reinforced concrete pavements contain steel mesh reinforcement (sometimes called distributed steel). In jointed reinforced concrete pavements, designers increase the joint spacing purposely, and include reinforcing steel to hold together intermediate cracks in each slab. The spacing between transverse joints is typically 30 feet (9.1 m) or more. In the past, some agencies used a spacing as great as 100 feet (30 m). During construction of the interstate system, most agencies in the Eastern and Midwestern United States laid jointed-reinforced pavement. Today only a handful of agencies employ this design, and its use is generally not recommended as both of the other types offer better performance and are easier to repair.
- Continuously reinforced concrete pavements do not require any transverse contraction joints. Transverse cracks are expected in the slab, usually at intervals of 3 to 5 ft (0.91 to 1.52 m). These pavements are designed with enough steel, 0.6–0.7% by cross-sectional area, so that cracks are held together tightly. Determining an appropriate spacing between the cracks is part of the design process for this type of pavement.

A concrete road in Ewing, New Jersey. The original pavement was laid in the 1950s and has not been significantly altered since.

Continuously reinforced designs may cost slightly more than jointed reinforced or jointed plain designs due to increased quantities of steel. Often the cost of the steel is offset by the reduced cost of concrete because a continuously reinforced design is nearly always significantly thinner than a jointed design for the same traffic loads. Properly designed, the two methods should demonstrate similar long-term performance and cost-effectiveness. A number of agencies have made policy decisions to use continuously reinforced designs in their heavy urban traffic corridors.

One of the major advantages of concrete pavements is they are typically stronger and more durable than asphalt roadways. They also can be grooved to provide a durable skid-resistant surface. A notable disadvantage is that they typically can have a higher initial cost, and can be more time-consuming to construct. This cost can typically be offset through the long life cycle of the pavement. Concrete pavement can be maintained over time utilizing a series of methods known as concrete pavement restoration which include diamond grinding, dowel bar retrofits, joint and crack sealing, cross-stitching, etc. Diamond grinding is also useful in reducing noise and restoring skid resistance in older concrete pavement.

The first street in the United States to be paved with concrete was Court Avenue in Bellefontaine, Ohio in 1893. The first mile of concrete pavement in the United States was on Woodward Avenue in Detroit, Michigan in 1909.

COMPOSITE PAVEMENT

An example of composite pavement: hot-mix asphalt overlaid onto Portland cement concrete pavement

Composite pavements combine a Portland cement concrete sublayer with an asphalt. They are usually used to rehabilitate existing roadways rather than in new construction.

Asphalt overlays are sometimes laid over distressed concrete to restore a smooth wearing surface. A disadvantage of this method is that movement in the joints between the underlying concrete slabs, whether from thermal expansion and contraction, or from deflection of the concrete slabs from truck axle loads, usually causes *reflective cracks* in the asphalt. To decrease reflective cracking, concrete pavement is broken apart through a *break and seat, crack and seat*, or *rubblization* process. Geosynthetics can be used for reflective crack control. With break and seat and crack and seat processes, a heavy weight is dropped on the concrete to induce cracking, then a heavy roller is used to seat the resultant pieces into the subbase. The main difference between the two processes is the equipment used to break the concrete pavement and the size of the resulting pieces.

The theory is frequent small cracks will spread thermal stress over a wider area than infrequent large joints, reducing the stress on the overlying asphalt pavement. Rubblization is a more complete fracturing of the old, worn-out concrete, effectively converting the old pavement into an aggregate base for a new asphalt road.

Whitetopping uses Portland cement concrete to resurface a distressed asphalt road.

Recycling

An asphalt milling machine in Boise, Idaho.

Distressed road materials can be reused when rehabilitating a roadway. The existing pavement is ground or broken up into small pieces, through a process called milling. It can then be transported to an asphalt or concrete plant and incorporated into new pavement, or recycled in place to form the base or subbase for new pavement. Some methods used include:

In-Place Recycling

- Rubblizing of concrete pavement. Existing concrete pavement is broken into gravel-sized particles. Any steel reinforcing is removed, then the remaining gravel-sized particles are compacted and overlaid with asphalt pavement.
- Cold in-place recycling. Bituminous pavement is ground or milled into small particles. The asphalt millings are blended with a small amount of asphalt emulsion or foamed bitumen, paved and compacted, allowed to cure for seven to ten days, then overlaid with asphalt.
- Hot in-place recycling. Bituminous pavement is heated to 250 to 300 °F (120 to 150 °C), milled, combined with a rejuvenating agent or virgin asphalt binder, and compacted. It may then be overlaid with a new asphalt overlay. This process only recycles the top two inches (50 mm) or less, so it can be used to correct rutting, polishing or other surface defects. It is not a good procedure for roads with structural failures. It also generates high heat and vapor emissions, and may not be a good candidate for built-up areas.

- Full depth reclamation is a process which pulverizes the full thickness of the asphalt pavement and some of the underlying material to provide a uniform blend of material. A binding agent may be mixed in to form a base course for the new pavement, or it may be left unbound to form a subbase course. Common binding agents include asphalt emulsion, fly ash, Portland cement or calcium chloride. It can also be mixed with aggregate, recycled asphalt millings, or crushed Portland cement to improve the gradation of the material, and can provide a design life cycle of 30 years with proper lab testing and field verification.

Bituminous Surface

Bituminous surface treatment (BST) or chipseal is used mainly on low-traffic roads, but also as a sealing coat to rejuvenate an asphalt concrete pavement. It generally consists of aggregate spread over a sprayed-on asphalt emulsion or cut-back asphalt cement. The aggregate is then embedded into the asphalt by rolling it, typically with a rubber-tired roller. This type of surface is described by a wide variety of regional terms including "chip seal", "tar and chip", "oil and stone", "seal coat", "sprayed seal" or "surface dressing" or as simply "bitumin."

BST is used on hundreds of miles of the Alaska Highway and other similar roadways in Alaska, the Yukon Territory, and northern British Columbia. The ease of application of BST is one reason for its popularity, but another is its flexibility, which is important when roadways are laid down over unstable terrain that thaws and softens in the spring.

Other types of BSTs include micropaving, slurry seals and Novachip. These are laid down using specialized and proprietary equipment. They are most often used in urban areas where the roughness and loose stone associated with chip seals is considered undesirable.

Thin Membrane Surface

A thin membrane surface (TMS) is an oil treated aggregate which is laid down upon a gravel road bed producing a dust free road. A TMS road reduces mud problems and provides stone free roads for local residents where loaded truck traffic is negligible. The TMS layer adds no significant structural strength, and so is used on secondary highways with low traffic volume and minimal weight loading. Construction involves minimal subgrade preparation, following by covering with a 50 to 100 millimetres (2.0–3.9 in) cold mix asphalt aggregate. The Operation Division of the Ministry of Highways and Infrastructure in Saskatchewan has the responsibility of maintaining 6,102 kilometres (3,792 mi) of thin membrane surface (TMS) highways.

Otta Seal

Otta seal is a low-cost road surface using a 16–30-millimetre (0.63–1.18 in) thick mixture of bitumen and crushed rock.

Gravel Surface

Gravel is known to have been used extensively in the construction of roads by soldiers of the Roman Empire (see Roman road) but a limestone-surfaced road, thought to date back to the Bronze Age, has been found in Britain. Applying gravel, or "**metalling,**" has had two distinct usages in road surfacing. The term ***road metal*** refers to the broken stone or cinders used in the construction or repair of roads or railways, and is derived from the Latin *metallum*, which means both "mine" and "quarry". The term originally referred to the process of creating a gravel roadway. The route of the roadway would first be dug down several feet and, depending on local conditions, French drains may or may not have been added. Next, large stones were placed and compacted, followed by successive layers of smaller stones, until the road surface was composed of small stones compacted into a hard, durable surface. "Road metal" later became the name of stone chippings mixed with tar to form the road surfacing material tarmac. A road of such material is called a "**metalled road**" in Britain, a "paved road" in Canada and the US, or a "sealed road" in parts of Canada, Australia and New Zealand.

A granular surface can be used with a traffic volume where the annual average daily traffic is 1,200 vehicles per day or less. There is some structural strength as the road surface combines a sub base and base and is topped with a double graded seal aggregate with emulsion. Besides the 4,929 kilometres (3,063 mi) of granular pavements maintained in Saskatchewan, around 40% of New Zealand roads are unbound granular pavement structures.

The decision whether to pave a gravel road or not often hinges on traffic volume. It has been found that maintenance costs for gravel roads often exceed the maintenance costs for paved or surface treated roads when traffic volumes exceed 200 vehicles per day.

Some communities are finding it makes sense to convert their low volume paved roads to aggregate surfaces.

Other Surfaces

Pavers (or **paviours**), generally in the form of pre-cast concrete blocks, are often used for aesthetic purposes, or sometimes at port facilities that see long-duration pavement loading. Pavers are rarely used in areas that see high-speed vehicle traffic.

Brick, cobblestone, sett, wood plank, and wood block pavements such as Nicolson pavement, were once common in urban areas throughout the world, but fell out of fashion in most countries, due to the high cost of labor required to lay and maintain them, and are typically only kept for historical or aesthetic reasons. In some countries, however, they are still common in local streets. In the Netherlands, brick paving has made somewhat of a comeback since the adoption of a major nationwide traffic safety program in 1997. From 1998 through 2007, more than 41,000 km of city streets were converted to local access roads with a speed limit of 30 km/h, for the purpose of traffic calming. One popular measure is to use brick paving - the noise and vibration slows motorists down. At the same time, it is not uncommon for cycle paths alongside a road to have a smoother surface than the road itself.

Likewise, macadam and tarmac pavements can still sometimes be found buried underneath asphalt concrete or Portland cement concrete pavements, but are rarely constructed today.

There are also other methods and materials to create pavements that have appearance of brick pavements. The first method to create brick texture is to heat an asphalt pavement and use metal wires to imprint a brick pattern using a compactor to create stamped asphalt. A similar method is to use rubber imprinting tools to press over a thin layer of cement to create decorative concrete. Another method is to use a brick pattern stencil and apply a surfacing material over the stencil. Materials that can be applied to give the color of the brick and skid resistance can be in many forms. An example is to use colored polymer-modified concrete slurry which can be applied by screeding or spraying. Another material is aggregate-reinforced thermoplastic which can be heat applied to the top layer of the brick-pattern surface. Other coating materials over stamped asphalt are paints and two-part epoxy coating.

Gallery

Concrete pavers

Replacing the old road with concrete blocks in Bo'ao Road area, Haikou City, Hainan, China

Polymer cement overlaying to change asphalt pavement to brick texture and color to create decorative crosswalk

ACOUSTICAL IMPLICATIONS

Roadway surfacing choices are known to affect the intensity and spectrum of sound emanating from the tire/surface interaction. Initial applications of this knowledge occurred in the early 1970s. Roadway surface types contribute differential noise effects of up to four dB, with chip seal type and grooved roads being the loudest and concrete surfaces without spacers being the quietest. Asphaltic surfaces perform intermediately relative to concrete and chip seal. These phenomena are, of course, highly influenced by vehicle speed. Rubberized asphalt has been shown to give a marginal 3–5 dB reduction in tire-pavement noise emissions, and a marginally discernible 1–3 decibel reduction in total road noise emissions when compared to conventional asphalt applications.

Cobbles

Rectangles

Decorative wavy pattern on La Rambla.

Decorative mock-brick pattern.

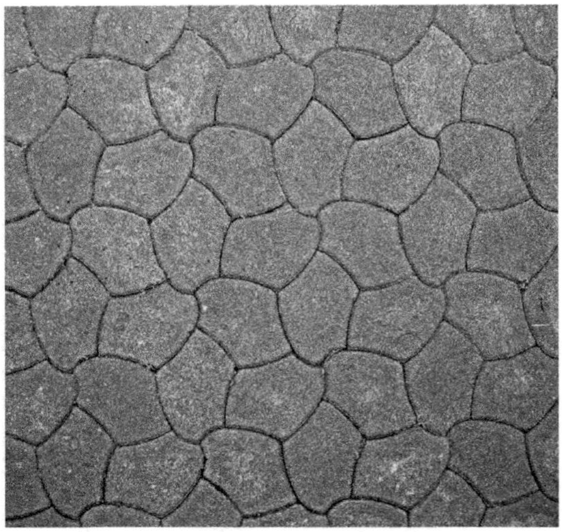

More decorative brickwork patterns.

Surface Deterioration

As pavement systems primarily fail due to fatigue (in a manner similar to metals), the damage done to pavement increases with the fourth power of the axle load of the vehicles traveling on it. According to the AASHO Road Test, heavily loaded trucks can do more than 10,000 times the damage done by a normal passenger car. Tax rates for trucks are higher than those for cars in most countries for this reason, though they are not levied in proportion to the damage done. Passenger cars are considered to have no practical effect on a pavement's service life, from a fatigue perspective.

Other failure modes include aging and surface abrasion. As years go by, the binder in a bituminous wearing course gets stiffer and less flexible. When it gets "old" enough, the surface will start losing aggregates, and macrotexture depth increases dramatically. If no maintenance action is done quickly on the wearing course, potholes will form. If the road is still structurally sound, a bituminous surface treatment, such as a chipseal or surface dressing can prolong the life of the road at low cost. In areas with cold climate, studded tires may be allowed on passenger cars. In Sweden and Finland, studded passenger car tires account for a very large share of pavement rutting.

Several design methods have been developed to determine the thickness and composition of road surfaces required to carry predicted traffic loads for a given period of time. Pavement design methods are continuously evolving. Among these are the Shell Pavement design method, and the American Association of State Highway and Transportation Officials (AASHTO) 1993 "Guide for Design of Pavement Structures". A new mechanistic-empirical design guide has been under development by NCHRP (Called Superpave Technology) since 1998. A new design guide called Mechanistic Empirical Pavement Design Guide (MEPDG) was developed and is about to be adopted by AASHTO.

The physical properties of a stretch of pavement can be tested using a falling weight deflectometer.

Further research by University College London into pavements has led to the development of an indoor, 80-sq-metre artificial pavement at a research centre called Pedestrian Accessibility and Movement Environment Laboratory (PAMELA). It is used to simulate everyday scenarios, from different pavement users to varying pavement conditions. There also exists a research facility near Auburn University, the NCAT Pavement Test Track, that is used to test experimental asphalt pavements for durability.

In addition to repair costs, the condition of a road surface has economic effects for road users. Rolling resistance increases on rough pavement, as does wear and tear of vehicle components. It has been estimated that poor road surfaces cost the average US driver $324 per year in vehicle repairs, or a total of $67 billion. Also, it has been estimated that small improvements in road surface conditions can decrease fuel consumption between 1.8 and 4.7%.

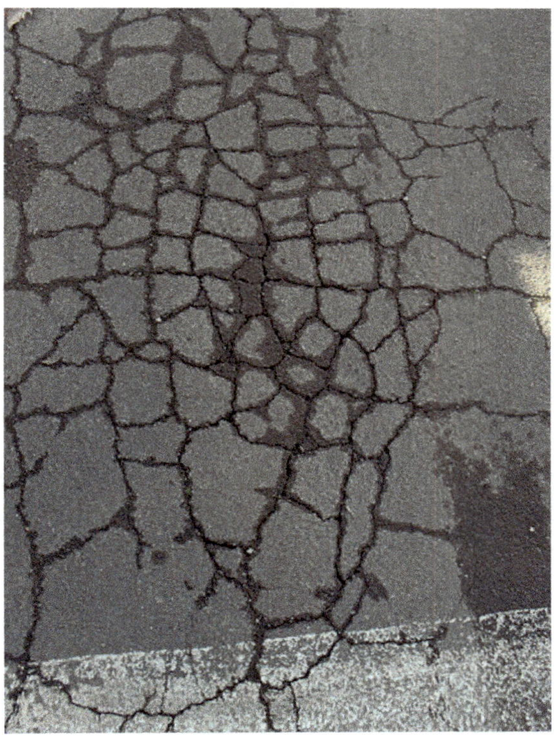

Deteriorating asphalt

Markings

Road surface markings are used on paved roadways to provide guidance and information to drivers and pedestrians. It can be in the form of mechanical markers such as cat's eyes, botts' dots and rumble strips, or non-mechanical markers such as paints, thermoplastic, plastic and epoxy.

REFERENCES

1. "NUAE Geosynthetics Ltd. News: Project Scout Moor Wind farm"(PDF). NUAE. May 2007. Retrieved 2 November 2008.
2. Anon (June 1991). "Highway construction/ Ground insulation"(PDF). Styropor: Technical information. Retrieved 29 January 2010.
3. "Warm Mix Asphalt Technologies and Research". Federal Highway Administration. 29 October 2008. Retrieved 4 August 2010.
4. "Hot, Warm, Luke Warm and Cold Mix Asphalt" (PDF). Cornell Local Roads Program. June 2009. Retrieved 4 August 2010.
5. Gerbrandt, Ron; Tim Makahoniuk; Cathy Lynn Borbely; Curtis Berthelot (2000). "Guidelines must be followed strictly – No exceptions" (PDF). Effect of Cold-in-place recycling on the Heavyweight Trucking Industry. 6th International Conference on Heavy Vehicle Weights and Dimension Proceedings. Retrieved 25 January 2009.
6. David Jones (2012). Advances in Pavement Design through Full-scale Accelerated Pavement Testing. CRC Press.
7. "Concrete Pavement Restoration" (PDF). Retrieved 7 April 2012.
8. "Concrete Pavement Rehabilitation Guide for Diamond Grinding". Retrieved 7 April 2012.
9. Ohio Historical Marker -
10. Lee, B. J.; Lee, H. (2004). "Position-Invariant Neural Network for Digital Pavement Crack Analysis". Computer-Aided Civil and Infrastructure Engineering. 19: 105–118. doi:10.1111/j.1467-8667.2004.00341.x.
11. Kulsea, Bill; Shawver, Tom; Kach, Carol (1980). Making Michigan Move: A History of Michigan Highways and the Michigan Department of Transportation. Lansing, Michigan: Michigan Department of Transportation. OCLC 8169232.
12. Khazanovich, L.; Lederle, R.; Tompkins, D.; Harvey, J.T.; Signore, J. (2012). "Guidelines for the Rehabilitation of Concrete Pavements Using Asphalt Overlays (FHWA TPF-5(149) Final Report)".
13. Moghadas Nejad, Fereidoon; Noory, Alireza; Toolabi, Saeed; Fallah, Shahab (8 August 2014). "Effect of using geosynthetics on reflective crack prevention". International Journal of Pavement Engineering. 16 (6): 477–487. doi:10.1080/10298436.2014.943128.
14. Lavin, Patrick (2003). Asphalt Pavements: A Practical Guide to Design, Production and Maintenance for Engineers and Architects. CRC Press. ISBN 9780203453292.
15. Rubblizing with Bituminous Concrete Overlay – 10 Years' Experience in Illinois, Illinois Department of Transportation, April 2002, http://www.dot.state.il.us/materials/research/pdf/137.pdf
16. "Asphalt Paving Principles". www.clrp.cornell.edu. Cornell Local Roads Program. March 2004. Retrieved 2016-10-05.

17. Budd, Godfrey. "An end to summer highway construction delays?". www.albertaconstructionmagazine.com. Retrieved 2016-10-05.

18. Sprayed Seal, Local Government & Municipal Knowledge Base, accessed 29 January 2010

19. Gransberg, Douglas D.; James, David M. B. (2005). Chip Seal Best Practices (Digitised online by Google books). National Cooperative Highway. Transportation Research Board. pp. 13–20. ISBN 9780309097444. Retrieved 25 February 2009.

20. Lazic, Zvjezdan; Ron Gerbrandt (2004). "Feasibility of Alternative salt storage structures in Saskatchewan Neilburg case study" (PDF). Measuring performance indicators for decision-making in winter maintenance operations. 2004 Annual conference of the Transportation Association of Canada. Saskatchewan Highways and Transportation. Archived from the original (pdf) on 18 March 2009. Retrieved 25 February 2009.

21. "Highways and Infrastructure — Government of Saskatchewan". Archived from the original on 8 February 2008. Retrieved 15 April 2008.

22. Manins, Rosie (28 February 2009). "New dust suppression method". Otago Daily Times. Retrieved 5 November 2011.

23. Anon (July 1998). "Bronze Age metalled road near Oxford". British Archaeology: News. Council for British Archaeology (36). Retrieved 29 January 2010.

24. Anon. "Road metal". Merriam-Webster online dictionary. Merriam Webster inc. Retrieved 29 January 2010.

25. Anon. "Metal". Online etymological dictionary. 2001 Douglas Harper. Retrieved 29 January 2010.

26. Anon. "Metalled Road". World Web Online. WordWeb Software. Retrieved 29 January 2010.

27. "Surfacing Aggregate" (PDF). Product brochure. Afrisam.com South Africa. 2008. Archived from the original (PDF) on 18 March 2009. Retrieved 25 January 2009.

28. Oeser, Markus; Sabine Werkmeister; Alvaro Gonzales; David Alabaster (2008). "Experimental and numerical simulation of loading impact on modified granular pavements" (PDF). 8th World Congress on Computational Mechanics 5th European Congress on computational methods in applied sciences and engineering ECCOMAS. 6th International Conference on Heavy Vehicle Weights and Dimension Proceedings. Retrieved 25 January 2009.

29. Mary C. Rukashaza-Mukome; et al. (2003). "Cost Comparison of Treatments Used to Maintain or Upgrade Aggregate Roads" (PDF). Proceedings of the 2003 Mid-Continent Transportation Research Symposium. Iowa State University. Retrieved 16 September 2011.

30. "Roads to Ruin: Towns Rip Up the Pavement". Wall Street Journal. 17 July 2010. Retrieved 16 September 2011.

31. "De balans opgemaakt: Duurzaam Veilig 1998–2007" [Sustainable Safety in the Netherlands - 1998–2007] (PDF) (in Dutch). SWOV, Dutch Institute for

Road Safety Research. 2009: 6 (English abstract). ISBN 978-90-73946-06-4. Retrieved 13 July 2014.

32. Hembrow, David (25 April 2011). "Road noise, cobbles and smooth asphalt". A view from the cycle path. Retrieved 14 August 2014.
33. Fred Young (23 February 2013). What can Seattle learn from Dutch street design? (presentation video). Seattle Neighborhood Greenways. Event occurs at 3:49 and 9:19 min. Retrieved 14 August 2014. ... the cycle track is asphalt (...) and the lane for cars is brick ...
34. "Endurablend Systems for Color Surfacing - Colors, Seals, Textures and Protects to Extend Service Life of Asphalt and Concrete Pavement Surfaces" (PDF). Retrieved 13 November 2014.
35. High-Performance Crosswalks (PDF). Alternative Paving Concepts. Retrieved 13 November 2014.
36. "Decorative Stamped Asphalt". ThermOTrack. Retrieved 13 November 2014.
37. C.M. Hogan, "Analysis of highway noise", Journal of Water, Air, & Soil Pollution, Volume 2, Number 3, Biomedical and Life Sciences and Earth and Environmental Science Issue, Pages 387–392, September, 1973, Springer Verlag, Netherlands ISSN 0049-6979
38. Statement Of Garth Dull For The Senate Epw Committee
39. "Scientists walk on tech pavement". BBC News. 12 September 2006. Retrieved 22 May 2010.
40. "The Value of Smooth". Better Roads. Randall Reilly. August 2011.

CHAPTER 6

Environmental Engineering

Chapter Contents

Environmental engineering is an incredibly broad field which is focused on using scientific principles to improve environmental conditions. Environmental engineers may use their skills to prevent damage to the environment, or to deal with the consequences of such damage. Many colleges and universities around the world offer training in environmental engineering to people who are interested in this field, and employment prospects are generally quite good.

INTRODUCTION

Environmental engineering is the branch of engineering concerned with the application of scientific and engineering principles for protection of human populations from the effects of adverse environmental factors; protection of environments, both local and global, from the potentially deleterious effects of natural and human activities; and improvement of environmental quality.

Environmental engineering can also be described as a branch of applied science and technology that addresses the issues of energy preservation, production asset and control of waste from human and animal activities. Furthermore, it is concerned with finding plausible solutions in the field of public health, such as waterborne diseases, implementing laws which promote adequate sanitation in urban, rural and recreational areas. It involves waste water management, air pollution control, recycling, waste disposal, radiation protection, industrial hygiene, environmental sustainability, public health and environmental engineering law. It also includes studies on the environmental impact of proposed construction projects.

Environmental engineers study the effect of technological advances on the environment. To do so, they conduct studies on hazardous-waste management to evaluate the significance of such hazards, advise on treatment and containment, and develop regulations to prevent mishaps. Environmental engineers design municipal water supply and industrial wastewater treatment systems. They address local and worldwide environmental issues such as the effects of acid rain, global warming, ozone depletion, water pollution and air pollution from automobile exhausts and industrial sources.

At many universities environmental engineering programs are offered at either the department of civil engineering or the department of chemical engineering at engineering faculties. Environmental "civil" engineers focus on hydrology, water resources management, bioremediation, and water treatment plant design. Environmental "chemical" engineers, on the other hand, focus on environmental chemistry, advanced air and water treatment technologies and separation processes.

Additionally, engineers are more frequently obtaining specialized training in law (J.D.) and are utilizing their technical expertise in the practices of environmental engineering law.

Most jurisdictions also impose licensing and registration requirements.

Development

Ever since people first recognized that their health is related to the quality of their environment, they have applied principles to attempt to improve the quality of their environment. The ancient Indian Harappan civilization utilized early sewers in some cities more than 5000 years ago. The Romans constructed aqueducts to prevent drought and to create a clean, healthful water supply for the metropolis of Rome. In the 15th century, Bavaria created laws restricting the development and degradation of alpine country that constituted the region's water supply.

The field emerged as a separate environmental discipline during the middle third of the 20th century in response to widespread public concern about water and pollution and increasingly extensive environmental quality degradation. However, its roots extend back to early efforts in public health engineering. Modern environmental engineering began in London in the mid-19th century when Joseph Bazalgette designed the first major sewerage system that reduced the incidence of waterborne diseases such as cholera. The introduction of drinking water treatment and sewage treatment in industrialized countries reduced waterborne diseases from leading causes of death to rarities.

In many cases, as societies grew, actions that were intended to achieve benefits for those societies had longer-term impacts which reduced other environmental qualities. One example is the widespread application of the pesticide DDT to control agricultural pests in the years following World War II. While the agricultural benefits were outstanding and crop yields increased dramatically thus reducing world hunger substantially, and malaria was controlled better than it ever had been, numerous species were brought to the verge of extinction due to the impact of the DDT on their reproductive cycles. The story of DDT as vividly told in Rachel Carson's *Silent Spring* (1962) is considered to be the birth of the modern environmental movement and of the modern field of "environmental engineering."

Conservation movements and laws restricting public actions that would harm the environment have been developed by various societies for millennia. Notable examples are the laws decreeing the construction of sewers in London and Paris in the 19th century and the creation of the U.S. national park system in the early 20th century.

Scope

Solid Waste Management

Environmental Impact Assessment and Mitigation

Scientists have air pollution dispersion models to evaluate the concentration of a pollutant at a receptor or the impact on overall air quality from vehicle exhausts and industrial flue gas stack emissions. To some extent, this field overlaps the desire to decrease carbon dioxide and other greenhouse gas emissions from combustion processes. They apply scientific and engineering principles to evaluate if there are likely to be any adverse impacts to water quality, air quality, habitat quality, flora and fauna, agricultural capacity, traffic impacts, social impacts, ecological impacts, noise impacts, visual (landscape) impacts, etc. If impacts are expected, they then develop mitigation measures to limit or prevent such impacts. An example of a mitigation measure would be the creation of wetlands in a nearby location to mitigate the filling in of wetlands necessary for a road development if it is not possible to reroute the road.

In the United States, the practice of environmental assessment was formally initiated on January 1, 1970, the effective date of the National Environmental Policy Act (NEPA). Since that time, more than 100 developing and developed nations either have planned specific analogous laws or have adopted procedure used elsewhere. NEPA is applicable to all federal agencies in the United States.

ENVIRONMENTAL IMPACT ASSESSMENT

Environmental assessment (EA) is the term used for the assessment of the environmental consequences (positive and negative) of a plan, policy, program, or concrete projects prior to the decision to move forward with the proposed action. In this context, the term "environmental impact assessment" (EIA) is usually used when applied to concrete projects by individuals or companies and the term "strategic environmental assessment" (SEA) applies to policies, plans and programmes most often proposed by organs of state (Fischer, 2016). Environmental assessments may be governed by rules of administrative procedure regarding public participation and documentation of decision making, and may be subject to judicial review.

The purpose of the assessment is to ensure that decision makers consider the environmental impacts when deciding whether or not to proceed with a project. The International Association for Impact Assessment (IAIA) defines an environmental impact assessment as "the process of identifying, predicting, evaluating and mitigating the biophysical, social, and other relevant effects of development proposals prior to major decisions being taken and commitments made." EIAs are unique in that they do not require adherence to a predetermined environmental outcome, but rather they require decision makers to account for environmental values in their decisions and to justify those decisions in light of detailed environmental studies and public comments on the potential environmental impacts.

CASI Global, New York (http://casiglobal.us/) The Global Certification body for CSR[expand acronym] & Sustainability goes one step further in this definition to include financial and supply chain effects. Any effects of the projects supply chain on the environment, ecology, human resources or country resources are to be included in the impact analysis. Most Global Fellows of CASI NY have to write a paper on the same and are authorized to undertake such projects in their country.

Engineering and consulting companies work hand in hand as contractors for mining, energy, oil & gas companies executing EIAs. Companies operating globally such as Arcadis, Royal HaskoningDHV, Golder Associates, Amec Foster Wheeler, Schlumberger Water Services (a Schlumberger company), ERM are an example of a much bigger pool of expertise globally. These contractors are the ones not only in charge of preparing an EIA study but most importantly getting these studies approved by each country government offices prior to the execution of a project. Each country will also have its own local contractors offering the same kind of service hence breaking out monopolies by increasing the supply of EIAs execution consultants.

Environmental impact assessments commenced in the 1960s, as part of increasing environmental awareness. EIAs involved a technical evaluation intended to contribute to more objective decision making. In the United States, environmental impact assessments obtained formal status in 1969, with enactment of the National Environmental Policy Act. EIAs have been used increasingly around the world. The number of "Environmental Assessments" filed every year "has vastly overtaken the number of more rigorous Environmental Impact Statements (EIS)." An Environmental Assessment is a "mini-EIS designed to provide sufficient information to allow the agency to decide whether the preparation of a full-blown Environmental Impact Statement (EIS) is necessary." EIA is an activity that is done to find out the impact that would be done before development will occur.

Methods

General and industry specific assessment methods are available including:

- *Industrial products* - Product environmental life cycle analysis (LCA) is used for identifying and measuring the impact of industrial products on the environment. These EIAs consider activities related to extraction of raw materials, ancillary materials, equipment; production, use, disposal and ancillary equipment.

- *Genetically modified plants* - Specific methods available to perform EIAs of genetically modified organisms include GMP-RAM and INOVA.
- *Fuzzy logic* - EIA methods need measurement data to estimate values of impact indicators. However, many of the environment impacts cannot be quantified, e.g. landscape quality, lifestyle quality and social acceptance. Instead information from similar EIAs, expert judgment and community sentiment are employed. Approximate reasoning methods known as fuzzy logic can be used. A fuzzy arithmetic approach has also been proposed and implemented using a software tool (TDEIA).

Follow-up

At the end of the project, an audit evaluates the accuracy of the EIA by comparing actual to predicted impacts. The objective is to make future EIAs more valid and effective. Two primary considerations are:

- *Scientific* - to examine the accuracy of predictions and explain errors
- *Management* - to assess the success of mitigation in reducing impacts

Audits can be performed either as a rigorous assessment of the null hypothesis or with a simpler approach comparing what actually occurred against the predictions in the EIA document.

After an EIA, the precautionary and polluter pays principles may be applied to decide whether to reject, modify or require strict liability or insurance coverage to a project, based on predicted harms.

The Hydropower Sustainability Assessment Protocol is a sector specific method for checking the quality of Environmental and Social assessments and management plans.

Around the World

Australia

The history of EIA in Australia could be linked to the enactment of the U.S. National Environment Policy Act (NEPA) in 1970, which made the preparation of environmental impact statements a requirement. In Australia, one might say that the EIA procedures were introduced at a State Level prior to that of the Commonwealth (Federal), with a majority of the states having divergent views to the Commonwealth. One of the pioneering states was New South Wales, whose State Pollution Control Commission issued EIA guidelines in 1974. At a Commonwealth (Federal) level, this was followed by passing of the Environment Protection (Impact of Proposals) Act in 1974. The Environment Protection and Biodiversity Conservation Act 1999 (EPBC) superseded the Environment Protection (Impact of Proposals) Act 1974 and is the current central piece for EIA in Australia on a Commonwealth (Federal) level. An important point to note is that this Commonwealth Act does not affect the validity of the States and Territories environmental and development assessments and approvals; rather the EPBC runs as a parallel to the

State/Territory Systems. Overlap between federal and state requirements is addressed via bilateral agreements or one off accreditation of state processes, as provided for in the EPBC Act.

The Commonwealth Level

The EPBC Act provides a legal framework to protect and manage nationally and internationally important flora, fauna, ecological communities and heritage places-defined in the EPBC Act as matters of 'national environmental significance'. Following are the eight matters of 'national environmental significance' to which the EPBC ACT applies:

- World Heritage sites
- National Heritage places
- RAMSAR wetlands of international significance
- Listed threatened species and ecological communities
- Migratory species protected under international agreements
- The Commonwealth marine environment
- Nuclear actions (including uranium mining)
- National Heritage.

In addition to this, the EPBC Act aims at providing a streamlined national assessment and approval process for activities. These activities could be by the Commonwealth, or its agents, anywhere in the world or activities on Commonwealth land; and activities that are listed as having a 'significant impact' on matters of 'national environment significance'.

The EPBC Act comes into play when a person (a 'proponent') wants an action (often called a 'proposal' or 'project') assessed for environmental impacts under the EPBC Act, he or she must refer the project to the Department of Environment, Water, Heritage and the Arts (Australia). This 'referral' is then released to the public, as well as relevant state, territory and Commonwealth ministers, for comment on whether the project is likely to have a significant impact on matters of national environmental significance. The Department of Environment, Water, Heritage and the Arts assess the process and makes recommendation to the minister or the delegate for the feasibility. The final discretion on the decision remains of the minister, which is not solely based on matters of 'national environmental significance' but also the consideration of social and economic impact of the project.

The Australian Government environment minister cannot intervene in a proposal if it has no significant impact on one of the eight matters of 'national environmental significance' despite the fact that there may be other undesirable environmental impacts. This is primarily due to the division of powers between the States and the Federal government and due to which the Australian Government environment minister cannot overturn a state decision.

There are strict civil and criminal penalties for the breach of EPBC Act. Depending on the kind of breach, civil penalty (maximum) may go up to $550,000 for an individual and $5.5 million for a body corporate, or for criminal penalty (maximum) of seven years imprisonment and/or penalty of $46,200.

The State and Territory Level

Australian Capital Territory (ACT)

EIA provisions within Ministerial Authorities in the ACT are found in the Chapters 7 and 8 of the *Planning and Development Act 2007* (ACT). EIA in ACT was previously administered with the help of Part 4 of the Land (Planning and Environment) Act 1991 (Land Act) and Territory Plan (plan for land-use). Note that some EIA may occur in the ACT on Commonwealth land under the EPBC Act (Cth). Further provisions of the *Australian Capital Territory (Planning and Land Management) Act 1988* (Cth) may also be applicable particularly to national land and "designated areas".

New South Wales (NSW)

In New South Wales, the Environment Planning Assessment Act 1979 (EPA) establishes three pathways for EIA. The first is under Part 5.1 of the EPAA, which provides for EIA of 'State Significant Infrastructure' projects. (From June 2011, this Part replaced Part 3A, which previously covered EIA of major projects). The second is under Part 4 of the Act dealing with development control. If a project does not require approval under Part 3A or Part 4 it is then potentially captured by the third pathway, Part 5 dealing with environment impact assessment.

Northern Territory (NT)

The EIA process in Northern Territory is chiefly administered under the Environmental Assessment Act (EAA). Although EAA is the primary tool for EIA in Northern Territory, there are further provisions for proposals in the Inquiries Act 1985 (NT).

Queensland (QLD)

There are four main EIA processes in Queensland. Firstly, under the Integrated Planning Act 1997 (IPA) for development projects other than mining. Secondly, under the Environmental Protection Act 1994 (EP Act) for some mining and petroleum activities. Thirdly, under the State Development and Public Works Organization Act 1971 (State Development Act) for 'significant projects'. Finally, Environment Protection and Biodiversity Conservation Act 1999 (Cth) for 'controlled actions'.

South Australia (SA)

The local governing tool for EIA in South Australia is the Development Act 1993. There are three levels of assessment possible under the Act in the form of an environment impact statement (EIS), a public environmental report (PER) or a Development Report (DR).

Tasmania (TAS)

In Tasmania, an integrated system of legislation is used to govern development and approval process, this system is a mixture of the Environmental Management and Pollution Control Act 1994 (EMPCA), Land Use Planning and Approvals Act 1993 (LUPAA), State Policies and Projects Act 1993 (SPPA), and Resource Management and Planning Appeals Tribunal Act 1993.

Victoria (VIC)

The EIA process in Victoria is intertwined with the Environment Effects Act 1978 and the Ministerial Guidelines for Assessment of Environmental Effects (made under the s. 10 of the EE Act).

Western Australia (WA)

The Environmental Protection Act 1986 (Part 4) provides the legislative framework for the EIA process in Western Australia.The EPA Act oversees the planning and development proposals and assesses their likely impacts on the environment.

Canada

In *Friends of the Oldman River Society v. Canada (Minister of Transportation)*,(SCC 1992) La Forest J of the Supreme Court of Canada described environmental impact assessment in terms of the proper scope of federal jurisdiction with respect to environments matters,

"Environmental impact assessment is, in its simplest form, a planning tool that is now generally regarded as an integral component of sound decision-making."

Supreme Court Justice La Forest cited (Cotton, Emond & 1981 245), "The basic concepts behind environmental assessment are simply stated: (1) early identification and evaluation of all potential environmental consequences of a proposed undertaking; (2) decision making that both guarantees the adequacy of this process and reconciles, to the greatest extent possible, the proponent's development desires with environmental protection and preservation."

La Forest referred to (Jeffrey 1989, 1.2,1.4) and (Emond 1978, p. 5) who described "...environmental assessments as a planning tool with both an information-gathering and a decision-making component" that provide "...an objective basis for granting or denying approval for a proposed development."

Justice La Forest addressed his concerns about the implications of Bill C-45 regarding public navigation rights on lakes and rivers that would contradict previous cases.(La Forest & 1973 178-80)

The Canadian Environmental Assessment Act 2012 (CEAA 2012) "and its regulations establish the legislative basis for the federal practice of environmental assessment in most regions of Canada." CEAA 2012 came into force July 6, 2012 and replaces the former *Canadian Environmental Assessment Act* (1995). EA is defined as a planning tool to identify, understand, assess and mitigate, where possible, the environmental effects of a project.

The purposes of this Act are: (a) to protect the components of the environment that are within the legislative authority of Parliament from significant adverse environmental effects caused by a designated project; (b) to ensure that designated projects that require the exercise of a power or performance of a duty or function by a federal authority under any Act of Parliament other than this Act to be carried out, are considered in a careful and precautionary manner to avoid significant adverse environmental effects; (c) to promote cooperation and coordinated action between federal and provincial governments with respect to environmental assessments; (d) to promote communication and cooperation with aboriginal peoples with respect to environmental assessments; (e) to ensure that opportunities are provided for meaningful public participation during an environmental assessment; (f) to ensure that an environmental assessment is completed in a timely manner; (g) to ensure that projects, as defined in section 66, that are to be carried out on federal lands, or those that are outside Canada and that are to be carried out or financially supported by a federal authority, are considered in a careful and precautionary manner to avoid significant adverse environmental effects; (h) to encourage federal authorities to take actions that promote sustainable development in order to achieve or maintain a healthy environment and a healthy economy; and (i) to encourage the study of the cumulative effects of physical activities in a region and the consideration of those study results in environmental assessments."

Opposition

Environmental Lawyer Dianne Saxe argued that the CEAA 2012 "allows the federal government to create mandatory timelines for assessments of even the largest and most important projects, regardless of public opposition." (Saxe 2012)

On 3 August 2012 the Canadian Environmental Assessment Agency nine "designated projects" with their timelines: Enbridge Northern Gateway Pipeline Joint Review Panel (JRP) 18 months; Marathon Platinum Group Metals and Copper Mine Project (JRP): 13 months; Site C Clean Energy Project (JRP) 8.5 months; Deep Geologic Repository Project (JRP) 17 months; Enbridge Northern Gateway Project (JRP) 18 months; Jackpine Mine Expansion Project (JRP) 11.5 months; Pierre River Mine Project: 8 months; New Prosperity Gold-Copper Mine Project (JRP) 7.5 months; Frontier Oil Sands Mine Project (JRP)8.5 months; EnCana/Cenovus Shallow Gas Infill Project (JRP) 5 months.

Saxe compares these timelines with environmental assessments for the Mackenzie Valley Pipeline. Thomas R. Berger, Royal Commissioner of the Mackenzie Valley Pipeline Inquiry (9 May 1977), worked extremely hard to ensure that industrial development on Aboriginal people's land resulted in benefits to those indigenous people.

On 22 April 2013, Official Opposition Environment critic Megan Leslie issued a statement claiming that the federal government's recent changes to "fish habitat protection, the Navigable Waters Protection Act and the Canadian Environmental Assessment Act", along with gutting existing laws and making cuts to science and research, "will be disastrous, not only for the environment, but also for Canadians' health and economic prosperity." On 26 September 2012, Leslie argued that with the changes to the Canadian Environmental Assessment Act that came into effect 6 July 2012, "seismic testing, dams, wind farms and power plants" no longer required any federal environmental assessment. She also claimed that because the CEAA 2012—which she claimed was rushed through Parliament—dismantled the CEAA 1995, the Oshawa ethanol plant project would no longer have a full federal environmental assessment. Mr. Peter Kent (Minister of the Environment) explained that the CEAA 2012 "provides for the Government of Canada and the Environmental Assessment Agency to focus on the large and most significant projects that are being proposed across the country." The 2,000 to 3,000-plus smaller screenings that were in effect under CEAA 1995 became the "responsibility of lower levels of government but are still subject to the same strict federal environmental laws." Anne Minh-Thu Quach, MP for Beauharnois—Salaberry, QC, argued that the mammoth budget bill dismantled 50 years of environmental protection without consulting Canadians about the "colossal changes they are making to environmental assessments." She claimed that the federal government is entering into "limited consultations, by invitation only, months after the damage was done."

China

The Environmental Impact Assessment Law (EIA Law) requires that an environmental impact assessment be completed prior to project construction. However, if a developer completely ignores this requirement and builds a project without submitting an environmental impact statement, the only penalty is that the environmental protection bureau (EPB) may require the developer to do a make-up environmental assessment. If the developer does not complete this make-up assessment within the designated time, only then is the EPB authorized to fine the developer. Even so, the possible fine is capped at a maximum of about US$25,000, a fraction of the overall cost of most major projects. The lack of more stringent enforcement mechanisms has resulted in a significant percentage of projects not completing legally required environmental impact assessments prior to construction.

China's State Environmental Protection Administration (SEPA) used the legislation to halt 30 projects in 2004, including three hydro-power plants under the Three Gorges Project Company. Although one month later (Note as a point of reference, that the typical EIA for a major project in the USA takes one to two years.), most of the 30 halted projects resumed their construction, reportedly having passed the environmental assessment, the fact that these key projects' construction was ever suspended was notable.

A joint investigation by SEPA and the Ministry of Land and Resources in 2004 showed that 30-40% of the mining construction projects went through the procedure of environment impact assessment as required, while in some areas only 6-7% did so. This partly explains why China has witnessed so many mining accidents in recent years.

SEPA alone cannot guarantee the full enforcement of environmental laws and regulations, observed Professor Wang Canfa, director of the centre to help environmental victims at China University of Political Science and Law. In fact, according to Wang, the rate of China's environmental laws and regulations that are actually enforced is estimated at barely 10%.

Egypt

Environmental Impact Assessment (EIA) EIA is implemented in Egypt under the umbrella of the Ministry of state for environmental affairs. The Egyptian Environmental Affairs Agency (EEAA) is responsible for the EIA services.

In June 1997, the responsibility of Egypt's first full-time Minister of State for Environmental Affairs was assigned as stated in the Presidential Decree no.275/1997. From thereon, the new ministry has focused, in close collaboration with the national and international development partners, on defining environmental policies, setting priorities and implementing initiatives within a context of sustainable development.

According to the Law 4/1994 for the Protection of the Environment, the Egyptian Environmental Affairs Agency (EEAA) was restructured with the new mandate to substitute the institution initially established in 1982. At the central level, EEAA represents the executive arm of the Ministry.

The purpose of EIA is to ensure the protection and conservation of the environment and natural resources including human health aspects against uncontrolled development. The long-term objective is to ensure a sustainable economic development that meets present needs without compromising future generations ability to meet their own needs. EIA is an important tool in the integrated environmental management approach.

EIA must be performed for new establishments or projects and for expansions or renovations of existing establishments according to the Law for the Environment.

EU

There is a wide range of instruments in the Environmental policy of the European Union. Among them the European Union has established a mix of mandatory and discretionary procedures to assess environmental impacts. European Union Directive (85/337/EEC) on Environmental Impact Assessments (known as the *EIA Directive*) was first introduced in 1985 and was amended in 1997. The directive was amended again in 2003, following EU signature of the 1998 Aarhus Convention, and once more in 2009. The initial Directive of 1985 and its three amendments have been codified in Directive 2011/92/EU of 13 December 2011. In 2001, the issue was enlarged to the assessment of plans and programmes by the so-called *Strategic Environmental Assessment (SEA) Directive* (2001/42/EC), which is now in force. Under the EU directive, an EIA must provide certain information to comply. There are seven key areas that are required:

- Description of the project
    - Description of actual project and site description
    - Break the project down into its key components, i.e. construction, operations, decommissioning
    - For each component list all of the sources of environmental disturbance
    - For each component all the inputs and outputs must be listed, e.g., air pollution, noise, hydrology
- Alternatives that have been considered

- Examine alternatives that have been considered
- Example: in a biomass power station, will the fuel be sourced locally or nationally?

■ Description of the environment

- List of all aspects of the environment that may be affected by the development
- Example: populations, fauna, flora, air, soil, water, humans, landscape, cultural heritage
- This section is best carried out with the help of local experts, e.g. the RSPB in the UK

■ Description of the significant effects on the environment

- The word significant is crucial here as the definition can vary
- 'Significant' must be defined
- The most frequent method used here is use of the Leopold matrix
- The matrix is a tool used in the systematic examination of potential interactions
- Example: in a windfarm development a significant impact may be collisions with birds

■ Mitigation

- This is where EIA is most useful
- Once section 4 is complete, it is obvious where impacts are greatest
- Using this information ways to avoid negative impacts should be developed
- Best working with the developer with this section as they know the project best
- Using the windfarm example again construction could be out of bird nesting seasons

■ Non-technical summary (EIS)

- The EIA is in the public domain and be used in the decision making process
- It is important that the information is available to the public
- This section is a summary that does not include jargon or complicated diagrams
- It should be understood by the informed lay-person

■ Lack of know-how/technical difficulties

- This section is to advise any areas of weakness in knowledge
- It can be used to focus areas of future research
- Some developers see the EIA as a starting block for poor environmental management

Annexed Projects

All projects are either classified as Annex 1 or Annex 2 projects. Those lying in Annex 1 are large scale developments such as motorways, chemical works, bridges, powerstations etc. These always require an EIA under the Environmental Impact Assessment Directive (85,337,EEC as amended). Annex 2 projects are smaller in scale than those referred to in Annex 1. Member States must determine whether these project shall be made subject to an assessment subject to a set of criteria set out in Annex 3 of codified Directive 2011/92/EU.

The Netherlands

EIA was implemented in Dutch legislation on September 1, 1987. The categories of projects that require an EIA are summarised in Dutch legislation, the Wet milieubeheer. The use of thresholds for activities makes sure that EIA is obligatory for those activities that may have considerable impacts on the environment.

For projects and plans that fit these criteria, an EIA report is required. The EIA report defines a.o. the proposed initiative, it makes clear the impact of that initiative on the environment and compares this with the impact of possible alternatives with less a negative impact.

Hong Kong

EIA in Hong Kong, since 1998, is regulated by the *Environmental Impact Assessment Ordinance 1997.*

The original proposal to construct the Lok Ma Chau Spur Line overground across the Long Valley failed to get through EIA, and the Kowloon–Canton Railway Corporation had to change its plan and build the railway underground. In April 2011, the EIA of the Hong Kong section of the Hong Kong-Zhuhai-Macau Bridge was found to have breached the ordinance, and was declared unlawful. The appeal by the government was allowed in September 2011. However, it was estimated that this EIA court case had increased the construction cost of the Hong Kong section of the bridge by HK$6.5 billion in money-of-the-day prices.

India

The Ministry of Environment, Forests and Climate Change (MoEFCC) of India has been in a great effort in Environmental Impact Assessment in India. The main laws in action are the Water Act(1974), the Indian Wildlife (Protection) Act (1972), the Air (Prevention and Control of Pollution) Act (1981) and the Environment (Protection) Act (1986),Biological Diversity Act(2002). The responsible body for this is the Central Pollution Control Board. Environmental Impact Assessment (EIA) studies need a significant amount of primary and secondary environmental data. Primary data are those collected in the field to define the status of the environment (like air quality data, water quality data etc.). Secondary data are those collected over the years that can be used to understand the existing environmental scenario of the study area. The environmental impact assessment

(EIA) studies are conducted over a short period of time and therefore the understanding of the environmental trends, based on a few months of primary data, has limitations. Ideally, the primary data must be considered along with the secondary data for complete understanding of the existing environmental status of the area. In many EIA studies, the secondary data needs could be as high as 80% of the total data requirement. EIC is the repository of one stop secondary data source for environmental impact assessment in India.

The Environmental Impact Assessment (EIA) experience in India indicates that the lack of timely availability of reliable and authentic environmental data has been a major bottle neck in achieving the full benefits of EIA. The environment being a multi-disciplinary subject, a multitude of agencies are involved in collection of environmental data. However, no single organization in India tracks available data from these agencies and makes it available in one place in a form required by environmental impact assessment practitioners. Further, environmental data is not available in enhanced forms that improve the quality of the EIA. This makes it harder and more time-consuming to generate environmental impact assessments and receive timely environmental clearances from regulators. With this background, the Environmental Information Centre (EIC) has been set up to serve as a professionally managed clearing house of environmental information that can be used by MoEF, project proponents, consultants, NGOs and other stakeholders involved in the process of environmental impact assessment in India. EIC caters to the need of creating and disseminating of organized environmental data for various developmental initiatives all over the country.

EIC stores data in GIS format and makes it available to all environmental impact assessment studies and to EIA stakeholders in a cost effective and timely manner. So that we can manage that in different proportions such as remedy measures etc.,

Korea, South

Recycling culture and policy Ministry of Environment

Malaysia

In Malaysia, Section 34A, Environmental Quality Act, 1974 requires developments that have significant impact to the environment are required to conduct the Environmental impact assessment.

Nepal

In Nepal, EIA has been integrated in major development projects since the early 1980s. In the planning history of Nepal, the sixth plan (1980–85), for the first time, recognized the need for EIA with the establishment of Environmental Impact Study Project (EISP) under the Department of Soil Conservation in 1982 to develop necessary instruments for integration of EIA in infrastructure development projects. However, the government of Nepal enunciated environment conservation related policies in the seventh plan (NPC, 1985–1990). To enforce this policy and make necessary arrangements, a series of guidelines were developed, thereby incorporating the elements of environmental factors right from the project formulation stage of the development plans and projects and to avoid or minimize adverse effects on the ecological system. In addition, it has also emphasized that EIAs of industry, tourism, water resources, transportation, urbanization, agriculture, forest and other developmental projects be conducted.

In Nepal, the government's Environmental Impact Assessment Guideline of 1993 inspired the enactment of the Environment Protection Act (EPA) of 1997 and the Environment Protection Rules (EPR) of 1997 (EPA and EPR have been enforced since 24 and 26 June 1997 respectively in Nepal) to internalizing the environmental assessment system. The process institutionalized the EIA process in development proposals and enactment, which makes the integration of IEE and EIA legally binding to the prescribed projects. The projects, requiring EIA or IEE, are included in Schedules 1 and 2 of the EPR, 1997 (GoN/MoLJPA 1997).

New Zealand

In New Zealand, EIA is usually referred to as *Assessment of Environmental Effects* (AEE). The first use of EIA's dates back to a Cabinet minute passed in 1974 called Environmental Protection and Enhancement Procedures. This had no legal force and only related to the activities of government departments. When the Resource Management Act was passed in 1991, an EIA was required as part of a resource consent application. Section 88 of the Act specifies that the AEE must include "such detail as corresponds with the scale and significance of the effects that the activity may have on the environment". While there is no duty to consult any person when making a resource consent application (Sections 36A and Schedule 4), proof of consultation is almost certain required by local councils when they decide whether or not to publicly notify the consent application under Section 93.

Russian Federation

As of 2004, the state authority responsible for conducting the State EIA in Russia has been split between two Federal bodies: 1) Federal service for monitoring the use of natural resources – a part of the Russian Ministry for Natural Resources and Environment and 2) Federal Service for Ecological, Technological and Nuclear Control. The two main pieces of environmental legislation in Russia are: The Federal Law 'On Ecological Expertise, 1995 and the 'Regulations on Assessment of Impact from Intended Business and Other Activity on Environment in the Russian Federation, 2000.

Federal Service for Monitoring the use of Natural Resources

In 2006, the parliament committee on ecology in conjunction with the Ministry for Natural Resources and Environment, created a working group to prepare a number of amendments to existing legislation to cover such topics as stringent project documentation for building of potentially environmentally damaging objects as well as building of projects on the territory of protected areas. There has been some success in this area, as evidenced from abandonment of plans to construct a gas pipe-line through the only remaining habitat of the critically endangered Amur leopard in the Russian Far East.

Federal Service for Ecological, Technological and Nuclear Control

The government's decision to hand over control over several important procedures, including state EIA in the field of all types of energy projects, to the Federal Service for Ecological, Technological and Nuclear Control had caused a major controversy and criticism from environmental groups that blamed the government for giving nuclear power industry control over the state EIA.

Not surprisingly the main problem concerning State EIA in Russia is the clear differentiation of jurisdiction between the two above-mentioned Federal bodies.

Sri Lanka

The National Environmental Act, 1998 requires environmental impact assessment for large scale projects in sensitive areas. It is enforced by the Central Environmental Authority.

United States

The National Environmental Policy Act of 1969 (NEPA), enacted in 1970, established a policy of environmental impact assessment for federal agency actions, federally funded activities or federally permitted/licensed activities that in the U. S. is termed "environmental review" or simply "the NEPA process." The law also created the Council on Environmental Quality, which promulgated regulations to codify the law's requirements. Under United States environmental law an Environmental Assessment (EA) is compiled to determine the need for an *Environmental Impact Statement* (EIS). Federal or federalized actions expected to subject or be subject to significant environmental impacts will publish a Notice of Intent to Prepare an EIS as soon as significance is known. Certain actions of federal agencies must be preceded by the NEPA process. Contrary to a widespread misconception, NEPA does not prohibit the federal government or its licensees/permittees from harming the environment, nor does it specify any penalty if an environmental impact assessment turns out to be inaccurate, intentionally or otherwise. NEPA requires that plausible statements as to the prospective impacts be disclosed in advance. The purpose of NEPA process is to ensure that the decision maker is fully informed of the environmental aspects and consequences prior to making the final decision.

Environmental Assessment

An **environmental assessment** (EA) is an environmental analysis prepared pursuant to the National Environmental Policy Act to determine whether a federal action would significantly affect the environment and thus require a more detailed *Environmental Impact Statement (EIS)*. The certified release of an Environmental Assessment results in either a *Finding of No Significant Impact (FONSI)* or an EIS.

The Council on Environmental Quality (CEQ), which oversees the administration of NEPA, issued regulations for implementing the NEPA in 1979. Eccleston reports that the NEPA regulations barely mention preparation of EAs. This is because the EA was originally intended to be a simple document used in relatively rare instances where an agency was not sure if the potential significance of an action would be sufficient to trigger preparation of an EIS. But today, because EISs are so much longer and complicated to prepare, federal agencies are going to great effort to avoid preparing EISs by using EAs, even in cases where the use of EAs may be inappropriate. The ratio of EAs that are being issued compared to EISs is about 100 to 1.

Likewise, even the preparation of an accurate EA is viewed today as an onerous burden by many entities responsible for the environmental review of a proposal. Federal agencies have responded by streamlining their regulations that implement NEPA environmental review, by defining categories of projects that by their well understood nature may be safely excluded from review under NEPA, and by drawing up lists of project types that have negligible material impact upon the environment and can thus be exempted.

Content

The Environmental Assessment is a concise public document prepared by the federal action agency that serves to:

- briefly provide sufficient evidence and analysis for determining whether to prepare an EIS or a Finding of No Significant Impact (FONSI)
- Demonstrate compliance with the act when no EIS is required
- facilitate the preparation of an EIS when a FONSI cannot be demonstrated

The Environmental Assessment includes a brief discussion of the purpose and need of the proposal and of its alternatives as required by NEPA 102(2)(E), and of the human environmental impacts resulting from and occurring to the proposed actions and alternatives considered practicable, plus a listing of studies conducted and agencies and stakeholders consulted to reach these conclusions. The action agency must approve an EA before it is made available to the public. The EA is made public through notices of availability by local, state, or regional clearing houses, often triggered by the purchase of a public notice advertisement in a newspaper of general circulation in the proposed activity area.

Structure

The structure of a generic Environmental Assessment is as follows:

- Summary
- Introduction
    - Background
    - Purpose and Need for Action
    - Proposed Action
    - Decision Framework
    - Public Involvement
    - Issues

- Alternatives, including the Proposed Action
    - Alternatives
    - Mitigation Common to All Alternatives
    - Comparison of Alternatives
- Environmental Consequences
- Consultation and Coordination

Procedure

The EA becomes a draft public document when notice of it is published, usually in a newspaper of general circulation in the area affected by the proposal. There is a 15-day review period required for an Environmental Assessment (30 days if exceptional circumstances) while the document is made available for public commentary, and a similar time for any objection to improper process. Commenting on the Draft EA is typically done in writing or email, submitted to the lead action agency as published in the notice of availability. An EA does not require a public hearing for verbal comments. Following the mandated public comment period, the lead action agency responds to any comments, and certifies either a FONSI or a Notice of Intent (NOI) to prepare an EIS in its public environmental review record. The preparation of an EIS then generates a similar but more lengthy, involved and expensive process.

Environmental Impact Statement

The adequacy of an environmental impact statement (EIS) can be challenged in federal court. Major proposed projects have been blocked because of an agency's failure to prepare an acceptable EIS. One prominent example was the Westway landfill and highway development in and along the Hudson River in New York City. Another prominent case involved the Sierra Club suing the Nevada Department of Transportation over its denial of the club's request to issue a supplemental EIS addressing air emissions of particulate matter and hazardous air pollutants in the case of widening U.S. Route 95 through Las Vegas. The case reached the United States Court of Appeals for the Ninth Circuit, which led to construction on the highway being halted until the court's final decision. The case was settled prior to the court's final decision.

Several state governments that have adopted "little NEPAs," state laws imposing EIS requirements for particular state actions. Some of those state laws such as the California Environmental Quality Act refer to the required environmental impact study as an environmental impact report.

This variety of state requirements produces voluminous data not just upon impacts of individual projects, but also in insufficiently researched scientific domains. For example, in a seemingly routine *Environmental Impact Report* for the city of Monterey, California, information came to light that led to the official federal endangered species listing of Hickman's potentilla, a rare coastal wildflower.

Transboundary Application

Environmental threats do not respect national borders. International pollution can have detrimental effects on the atmosphere, oceans, rivers, aquifers, farmland, the weather and biodiversity. Global climate change is transnational. Specific pollution threats include acid rain, radioactive contamination, debris in outer space, stratospheric ozone depletion and toxic oil spills. The Chernobyl disaster, precipitated by a nuclear accident on April 26, 1986, is a stark reminder of the devastating effects of transboundary nuclear pollution.

Environmental protection is inherently a cross-border issue and has led to the creation of transnational regulation via multilateral and bilateral treaties. The United Nations Conference on the Human Environment (UNCHE or Stockholm Conference) held in Stockholm in 1972 and the United Nations Conference on the Environment and Development (UNCED or Rio Summit, Rio Conference, or Earth Summit) held in Rio de Janeiro in 1992 were key in the creation of about 1,000 international instruments that include at least some provisions related to the environment and its protection.

The United Nations Economic Commission for Europe's Convention on Environmental Impact Assessment in a Transboundary Context was negotiated to provide an international legal framework for transboundary EIA.

However, as there is no universal legislature or administration with a comprehensive mandate, most international treaties exist parallel to one another and are further developed without the benefit of consideration being given to potential conflicts with other agreements. There is also the issue of international enforcement. This has led to duplications and failures, in part due to an inability to enforce agreements. An example is the failure of many international fisheries regimes to restrict harvesting practises.

Criticism

As per Jay *et al.*, EIA is used as a decision aiding tool rather than decision making tool. There is growing dissent about them as their influence on decisions is limited. Improved training for practitioners, guidance on best practice and continuing research have all been proposed.

EIAs have been criticized for excessively limiting their scope in space and time. No accepted procedure exists for determining such boundaries. The boundary refers to 'the spatial and temporal boundary of the proposal's effects'. This boundary is determined by the applicant and the lead assessor, but in practice, almost all EIAs address only direct and immediate on-site effects.

Development causes both direct and indirect effects. Consumption of goods and services, production, use and disposal of building materials and machinery, additional land use for activities of manufacturing and services, mining and refining, etc., all have environmental impacts. The indirect effects of development can be much higher than the direct effects examined by an EIA. Proposals such as airports or shipyards cause wide-ranging national and international effects, which should be covered in EIAs.

Broadening the scope of EIA can benefit the conservation of threatened species. Instead of concentrating on the project site, some EIAs employed a habitat-based approach that focused on much broader relationships among humans and the environment. As a result, alternatives that reduce the negative effects to the population of whole species, rather than local subpopulations, can be assessed.

Thissen and Agusdinata have argued that little attention is given to the systematic identification and assessment of uncertainties in environmental studies which is critical in situations where uncertainty cannot be easily reduced by doing more research. In line with this, Maier et al. have concluded on the need to consider uncertainty at all stages of the decision-making process. In such a way decisions can be made with confidence or known uncertainty. These proposals are justified on data that shows that environmental assessments fail to predict accurately the impacts observed. Tenney et al. and Wood et al. have reported evidence of the intrinsic uncertainty attached to EIAs predictions from a number of case studies worldwide. The gathered evidence consisted of comparisons between predictions in EIAs and the impacts measured during, or following project implementation. In explaining this trend, Tenney et al. have highlighted major causes such as project changes, modelling errors, errors in data and assumptions taken and bias introduced by people in the projects analyzed. Cardenas and Halman provide a comprehensive review on the issues of uncertainty in environmental impact assessments.

Water Supply and Treatment

Sewage treatment plant, Australia

They evaluate the water balance within a watershed and determine the available water supply, the water needed for various needs in that watershed, the seasonal cycles of water movement through the watershed and they develop systems to store, treat, and convey water for various uses. Water is treated to achieve water quality objectives for the end uses. In the case of a potable water supply, water is treated to minimize the risk of infectious disease transmission, the risk of non-infectious illness, and to create a palatable water flavor. Water distribution systems are designed and built to provide adequate water pressure and flow rates to meet various end-user needs such as domestic use, fire suppression, and irrigation.

Wastewater Treatment

Water pollution

There are numerous wastewater treatment technologies. A wastewater treatment train can consist of a primary clarifier system to remove solid and floating materials, a secondary treatment system consisting of an aeration basin followed by flocculation and sedimentation or an activated sludge system and a secondary clarifier, a tertiary biological nitrogen removal system, and a final disinfection process. The aeration basin/ activated sludge system removes organic material by growing bacteria (activated sludge). The secondary clarifier removes the activated sludge from the water. The tertiary system, although not always included due to costs, is becoming more prevalent to remove nitrogen and phosphorus and to disinfect the water before discharge to a surface water stream or ocean outfall.

Air Pollution Management

Scientists have developed air pollution dispersion models to evaluate the concentration of a pollutant at a receptor or the impact on overall air quality from vehicle exhausts and industrial flue gas stack emissions. To some extent, this field overlaps the desire to decrease carbon dioxide and other greenhouse gas emissions from combustion processes.

Environmental Protection Agency

The U.S. Environmental Protection Agency (EPA) is one of the many agencies that work with environmental engineers to solve key issues. An important component of EPA's mission is to protect and improve air, water, and overall environmental quality in order to avoid or mitigate the consequences of harmful effects.

Ecological engineering for sustainable agriculture in arid and semiarid West African regions

Ecological engineering offers new alternatives for the management of agricultural systems that are more tailored to the ever-changing social and environmental necessities in these regions. This requires managing the complexity of agrosystems, while striving to mimic the functioning of natural ecosystems of West African drylands and taking advantage of traditional practices and local know-how resulting from a long process of adaptation to environmental constraints.

- Acting on biodiversity. Biodiversity is essential to the productivity of ecosystems and their temporal stability under the impact of external disturbances. Several ecological processes related to biodiversity may be intensified for the benefit of agrosilvopastoral systems: promoting diversity and soil microorganism activity to benefit plants, associating and utilizing the mutual benefits of plants
- Utilizing organic matter and nutrient cycles. The productivity of agrosystems with low chemical input use in dryland regions is primarily based on efficient organic resource management, and in turn on the nutrient and energy flows they induce. It is thus possible to intervene at several levels: enhancing crop-livestock farming integration to preserve natural resources, restoring the biological activity of soils via specific organic inputs, supplying nutrients to plants locally.
- Enhancing available water use. Water supplies are limited and irregular in dryland areas. Current management of these supplies—which involves capturing rainwater and surface runoff—could be improved in several ways: adapting to erratic rainfall or drought risks by focusing on: (i) the organization of the farm and community (farm plot patterns in association with the random rainfall distribution, etc.), and on (ii) cropping techniques to reduce crop water needs (plant choices, weeding, etc.), preserving water in crop fields by hampering runoff, accounting for the essential role of trees regarding soil and water in drylands.
- Managing landscapes and associated ecological processes. Ecological crop pest regulation by their natural enemies is one ecosystem service provided by biodiversity. Better pest management could be considered in association with promoting biodiversity at different scales, e.g. from the plant to the landscape.

Education

Courses aimed at developing graduates with specific skills in environmental systems or environmental technology are becoming more common and fall into broad classes:

- *Mechanical engineering* courses oriented towards designing machines and mechanical systems for environmental use such as water treatment facilities, pumping stations, garbage segregation plants and other mechanical facilities;
- *Environmental engineering or environmental systems* courses oriented towards a civil engineering approach in which structures and the landscape are constructed to blend with or protect the environment;

- *Environmental chemistry*, *sustainable chemistry* or *environmental chemical engineering* courses oriented towards understanding the effects (good and bad) of chemicals in the environment. Focus on mining processes, pollutants and commonly also cover biochemical processes;
- *Environmental technology* courses oriented towards producing electronic or electrical graduates capable of developing devices and artifacts able to monitor, measure, model and control environmental impact, including monitoring and managing energy generation from renewable sources.

Prominent Environmental Engineers

- Robert A. Gearheart
- Paul V. Roberts
- Daniel A. Vallero
- Abel Wolman

REFERENCES

1. The American Academy of Environmental Engineers
2. Beychok, Milton R. (1967). Aqueous Wastes from Petroleum and Petrochemical Plants (1st ed.). John Wiley & Sons. LCCN 67019834.
3. Tchobanoglous, G.; Burton, F.L. & Stensel, H.D. (2003). Wastewater Engineering (Treatment Disposal Reuse) / Metcalf & Eddy, Inc. (4th ed.). McGraw-Hill Book Company. ISBN 0-07-041878-0.
4. Turner, D.B. (1994). Workbook of atmospheric dispersion estimates: an introduction to dispersion modeling (2nd ed.). CRC Press. ISBN 1-56670-023-X.
5. Beychok, M.R. (2005). Fundamentals Of Stack Gas Dispersion (4th ed.). author-published. ISBN 0-9644588-0-2.
6. “Architecture and Engineering Occupations : Occupational Outlook Handbook : U.S. Bureau of Labor Statistics”. Bls.gov. 2012-03-29. Retrieved 2013-07-01.
7. Career Information Center. Agribusiness, Environment, and Natural Resources (9th ed.). Macmillan Reference. 2007.
8. “Funding - Environmental Engineering - US National Science Foundation (NSF)”. nsf.gov. Retrieved 2013-07-01.
9. Sustainable Development (n.d.) Environmental Science. Detroit. 2009.
10. McGraw-Hill Encyclopedia of Environmental Science and Engineering (3rd ed.). McGraw-Hill, Inc. 1993.
11. Sims, J. (2003). Activated sludge, Environmental Encyclopedia. Detroit.
12. D. Masse; JL. Chotte; E. Scopel (2015). “Ecological engineering for sustainable agriculture in arid and semiarid West African regions”. Fiche thématique du CSFD (11): 2.

13. "Principle of Environmental Impact Assessment Best Practice" (PDF). International Association for Impact Assessment. 1999.
14. Holder, J., (2004), Environmental Assessment: The Regulation of Decision Making, Oxford University Press, New York; For a comparative discussion of the elements of various domestic EIA systems, see Christopher Wood Environmental Impact Assessment: A Comparative Review (2 ed, Prentice Hall, Harlow, 2002).
15. "EIA engineering and consulting companies". Digiscend.com. 2015. Retrieved 2015-08-01.
16. Clark & Canter 1997, p. 199.
17. Rychlak & Case 2010, p. 111-120.
18. Kershner 2011.
19. Daniel, S., Tsoulfas, G., Pappis, C., & Rachaniotis, N. (2004) Aggregating and evaluating the results of different Environmental Impact Assessment methods Ecological indicators 4:125-138
20. Hitzschky, K., & Silviera, J. (2009) A proposed impact assessment method for genetically modified plants (As-GMP method) Environmental Impact Assessment review 29: 348-368
21. Peche, R., & Rodriguez, E., (2009) Environmental impact Assessment procedure: A new approach based on Fuzzy logic Environmental Impact Assessment review 29:275-283
22. Duarte O. (2000) Técnicas Difusas en la Evaluación de Impacto Ambiental. Ph.D. Thesis, Universidad de Granada
23. ARAI Research Group, Granada, Spain. "Environmental Impact Assessment at University of Granada."
24. Wilson, L., (1998), A Practical Method for Environmental Impact Assessment Audits Environ Impact Assess Rev 18: 59-71
25. Elliott, M. & Thomas, I. (2009), "Environment Impact Assessment in Australia: Theory and Practice, 5th Edn, Federation Press, Sydney"
26. The Environment Protection and Biodiversity Conservation Act, Australia: The Department of the Environment, Water, Heritage and the Arts, retrieved 9 September 2010
27. The Northern Territory Government, viewed 10 September 2010,
28. The Environment Defenders, viewed 10 September 2010, EDO factsheet
29. The Law Handbook, viewed 9 September 2010,
30. The Government of Western Australia, viewed 9 September 2010,
31. SCC 1992.
32. Roger, Cotton; Emond, D. Paul (1981). John Swaigen, ed. Environmental Impact Assessment. Environmental Rights in Canada. Toronto, Ontario: Butterworths.
33. Jeffery, Michael I. (1989). Environmental Approvals in Canada. Toronto, Ontario: Butterworths.

34. Emond, D. P. (1978). Environmental Assessment Law in Canada. Toronto, Ontario: Emond Montgomery Ltd.
35. La Forest,, Gérard V. (1973). Water Law in Canada. Ottawa: Information Canada.
36. {{{year}}} SCC {{{num}}} (18 July 2012)
37. 2012 SCC {{{num}}}
38. {{{year}}} SCC {{{num}}}
39. CEAA (2012). "What is the Canadian Environmental Assessment Act, 2012?". Basics of Environmental Assessment. Canadian Environmental Assessment Agency.
40. CEAA 2012.
41. Saxe, =Dianne (9 August 2012). "Federal environmental assessments will be rushed". Environmental Law and Litigation.
42. Saxe 2012.
43. Host:Allan McFee (9 May 1977). "The Berger Report is released". As It Happens. transcript. Toronto. CBC Radio. CBC Radio 1. Retrieved 9 December 2011.
44. "Statement by Official Opposition Environment critic Megan Leslie on Earth Day". NDP. 22 April 2013.
45. Megan Leslie ({{{year}}}). Question period
46. Wang, Alex (2007-02-05). "Environmental protection in China: the role of law".
47. Gu, Lin (2005-09-29). "China Improves Enforcement of Environmental Laws". China Features.
48. "About MSEA". EEAA. Retrieved 2013-01-03.
49. Watson, Michael (November 13–15, 2003). "Environmental Impact Assessment and European Community Law". XIV International Conference "Danube-River of Cooperation".
50. Council Directive 85/337/EEC on the Assessment of the Effects of Certain Public and Private Projects on the Environment (1985-06-27) from Eur-Lex
51. DIRECTIVE 2011/92/EU OF THE EUROPEAN PARLIAMENT AND OF THE COUNCIL of 13 December 2011 on the assessment of the effects of certain public and private projects on the environment from Eur-Lex
52. Directive 2001/42/EC of the European Parliament and of the Council (2001-06-27) from Eur-Lex
53. List of criteria for EIA/SEA
54. "LCQ1: Judicial review case regarding the Environmental Impact Assessment reports of the Hong Kong-Zhuhai-Macao Bridge". Press Releases. www.info.gov.hk. October 26, 2011. Retrieved October 27, 2011.
55. Shibani Ghosh, "Demystifying the Environmental Clearance Process in India", NUJS LAW REVIEW, January 2, 2015
56. http://www.doe.gov.my/portal/wp-content/uploads/2010/07/Appendix_1.pdf
57. Department of Environmental Protection, Russia

58. Central Environmental Authority. "Environmental Impact Assessment (EIA) Procedure in Sri Lanka". Central Environmental Authority. Central Environmental Authority. Retrieved 1 November 2016.

59. United States. National Environmental Policy Act, P.L. 91-190, 83 Stat. 852, 42 U.S.C. § 4321 et seq. Approved January 1, 1970.

60. U.S. Council on Environmental Quality. "NEPA and Agency Planning." Code of Federal Regulations, 40 C.F.R. 1501.

61. Eccleston, Charles; Doub, J. Peyton (2012). Preparing NEPA Environmental Assessments: A User's Guide to Best Professional Practices. CRC Press. ISBN 9781439808825.

62. Sierra Club v. United States Army Corps of Engineers,701 F.2d 1011, 18 ERC 1748. (2d Cir., 02/25/1983)

63. Ritter, John (2003-06-03). "Lawsuit pits risks and roads". USA Today.

64. "Sive,D. & Chertok,M., "Little NEPAs" and Environmental Impact Assessment Procedures" (PDF). Retrieved 2013-01-03.

65. Rogers, Mark. Endangered Flora of California. lulu.com. p. 106. ISBN 978-1312862784. Retrieved November 25, 2016.

66. C.M. Hogan, G. Deghi, M. Papineau et al., Environmental Impact Report for the Pebble Beach Properties project by Del Monte Forest, Earth Metrics Inc. Prepared for the city of Monterey and State of California Clearinghouse (1992)

67. U.S. Federal Register: August 2, 1995 (Volume 60, Number 148, Pages 39326-39337)

68. Sands, P., (1989), The Environment, Community and International Law, Harvard International Law Lournal, 393, p402

69. Weiss, E., (1999), Understanding Compliance with International Environmental Agreements: The Bakers Dozen Myths, Univ Richmond L.R. 32, 1555

70. "Convention on Environmental Impact Assessment in a Transboundary Context (Espoo, 1991)". Unece.org. Retrieved 2013-01-03.

71. Wolfrum, R., & Matz, N., (2003), Conflicts in International Environmental Law, Max-Planck-Institut für Ausländisches Öffentliches Recht und Völkerrech

72. Young, O., (1999), The Effectiveness of International Environmental Regimes, MIT Press

73. Jay, S.; Jones, C.; Slinn, P.; Wood, C. (2007). "Environmental Impact Assessment: Retrospect and Prospect". Environmental Impact Assessment Review. Elsevier. 27 (4): 289–300. doi:10.1016/j.eiar.2006.12.001.

74. Lenzen, M.; Murray, S.; Korte, B.; Dey, C. (2003). "Environmental impact assessment including indirect effects—a case study using input-output analysis". Environmental Impact Assessment Review. Elsevier. 23 (3): 263–282. doi:10.1016/S0195-9255(02)00104-X.

75. Shepherd, A.; Ortolano, L. (1996). "Strategic environmental assessment for sustainable urban development". Environmental Impact Assessment Review.

Elsevier. 16 (4-6): 321–335. doi:10.1016/S0195-9255(96)00071-6.

76. Fernandes, João P. (2000). "EIA procedure, Landscape ecology and conservation management—Evaluation of alternatives in a highway EIA process". Environmental Impact Assessment Review. Elsevier. 20 (6): 665–680. doi:10.1016/S0195-9255(00)00060-3.

77. Thissen, WIH; Agusdinata, DB (2008). "Handling deep uncertainties in impact assessment". In proceedings of the 28th annual conference of the IAIA May 5–9, 2008, Perth, Australia. IAIA.

78. Maier, HR; Ascough, JC; Wattenbach, M; Renschler, CS; Labiosa, WB (2008). "Uncertainty in Environmental Decision Making: Issues, Challenges and Future Directions". Publications from USDA-ARS/UNL Faculty. Paper 399. USDA-ARS/UNL Faculty.

79. Tenney, A; Kværner, J; Gjerstad, KI (2006). "Uncertainty in environmental impact assessment predictions: the need for better communication and more transparency". Impact Assessment and Project Appraisal. 24 (1): 45–56.

80. Wood, C; , Dipper, B; Jones, C (2000). "Auditing the assessments of the environmental impacts of planning projects". Journal of Environmental Planning and Management. 43 (1): 23–47.

81. Cardenas, IC; Halman, JIM. "Coping with uncertainty in environmental impact assessments. Open techniques". Environmental Impact Assessment Review. Elsevier. 60: 24–39. doi:10.1016/j.eiar.2016.02.006.

Municipal or Urban Engineering

Chapter Contents

- **Introduction**
- **Water Supply Network**
- **Sanitary Sewer**
- **Urban Design**
- **Urban Planning**

Municipal engineering is concerned with municipal infrastructure. This involves specifying, designing, constructing, and maintaining streets, sidewalks, water supply networks, sewers, street lighting, municipal solid waste management and disposal, storage depots for various bulk materials used for maintenance and public works (salt, sand, etc.), public parks and bicycle paths. In the case of underground utility networks, it may also include the civil portion (conduits and access chambers) of the local distribution networks of electrical and telecommunications services. It can also include the optimizing of waste collection and bus service networks. Some of these disciplines overlap with other civil engineering specialties, however municipal engineering focuses on the coordination of these infrastructure networks and services, as they are often built simultaneously, and managed by the same municipal authority.

INTRODUCTION

Municipal or urban engineering applies the tools of science, art and engineering in an urban environment.

Summary

Municipal engineering is concerned with municipal infrastructure. This involves specifying, designing, constructing, and maintaining streets, sidewalks, water supply networks, sewers, street lighting, municipal solid waste management and disposal, storage depots for various bulk materials used for maintenance and public works (salt, sand, etc.), public parks and cycling infrastructure. In the case of underground utility networks, it may also include the civil portion (conduits and access chambers) of the local distribution networks of electrical and telecommunications services. It can also include the optimizing of garbage collection and bus service networks. Some of these disciplines overlap with other civil engineering specialties, however municipal engineering focuses on the coordination of these infrastructure networks and services, as they are often built simultaneously (for a given street or development project), and managed by the same municipal authority.

WATER SUPPLY NETWORK

A water supply system or water supply network is a system of engineered hydrologic and hydraulic components which provide water supply. A water supply system typically includes:

- A drainage basin (see water purification - sources of drinking water).
- A raw water collection point (above or below ground) where the water accumulates, such as a lake, a river, or groundwater from an underground aquifer. Raw water may be transferred using uncovered ground-level aqueducts, covered tunnels or underground water pipes to water purification facilities.
- Water purification facilities. Treated water is transferred using water pipes (usually underground).
- Water storage facilities such as reservoirs, water tanks, or water towers. Smaller water systems may store the water in cisterns or pressure vessels. Tall buildings may also need to store water locally in pressure vessels in order for the water to reach the upper floors.
- Additional water pressurizing components such as pumping stations may need to be situated at the outlet of underground or above ground reservoirs or cisterns (if gravity flow is impractical).
- A pipe network for distribution of water to the consumers (which may be private houses or industrial, commercial or institution establishments) and other usage points (such as fire hydrants).
- Connections to the sewers (underground pipes, or aboveground ditches in some developing countries) are generally found downstream of the water consumers, but the sewer system is considered to be a separate system, rather than part of the water supply system.

Water Abstraction and Raw Water Transfer

Raw water (untreated) is collected from a surface water source (such as an intake on a lake or a river) or from a groundwater source (such as a water well drawing from an underground aquifer) within the watershed that provides the water resource.

The raw water is transferred to the water purification facilities using uncovered aqueducts, covered tunnels or underground water pipes.

Water Treatment

Virtually all large systems must treat the water; a fact that is tightly regulated by global, state and federal agencies, such as the World Health Organization (WHO) or the United States Environmental Protection Agency (EPA). Water treatment must occur before the product reaches the consumer and afterwards (when it is discharged again). Water purification usually occurs close to the final delivery points to reduce pumping costs and the chances of the water becoming contaminated after treatment.

Traditional surface water treatment plants generally consists of three steps: clarification, filtration and disinfection. Clarification refers to the separation of particles (dirt, organic matter, etc.) from the water stream. Chemical addition (i.e. alum, ferric chloride) destabilizes the particle charges and prepares them for clarification either by settling or floating out of the water stream. Sand, anthracite or activated carbon filters refine the water stream, removing smaller particulate matter. While other methods of disinfection exist, the preferred method is via chlorine addition. Chlorine effectively kills bacteria and most viruses and maintains a residual to protect the water supply through the supply network.

Water Distribution Network

The Central Arizona Project Aqueduct transfers untreated water

Most (treated) water distribution happens through underground pipes

Pressurizing the water is required between the small water reserve and the end-user

The product, delivered to the point of consumption, is called potable water if it meets the water quality standards required for human consumption.

The water in the supply network is maintained at positive pressure to ensure that water reaches all parts of the network, that a sufficient flow is available at every take-off point and to ensure that untreated water in the ground cannot enter the network. The water is typically pressurised by pumps that pump water in to storage tanks constructed at the highest local point in the network. One network may have several such service reservoirs .

In small domestic systems, the water may be pressurised by a pressure vessel or even by an underground cistern (the latter however does need additional pressurizing). This eliminates the need of a water-tower or any other heightened water reserve to supply the water pressure.

These systems are usually owned and maintained by local governments, such as cities, or other public entities, but are occasionally operated by a commercial enterprise (see water privatization). Water supply networks are part of the master planning of communities, counties, and municipalities. Their planning and design requires the expertise of city planners and civil engineers, who must consider many factors, such as location, current demand, future growth, leakage, pressure, pipe size, pressure loss, fire fighting flows, etc. — using pipe network analysis and other tools.

As water passes through the distribution system, the water quality can degrade by chemical reactions and biological processes. Corrosion of metal pipe materials in the distribution system can cause the release of metals into the water with undesirable aesthetic and health effects. Release of iron from unlined iron pipes can result in customer reports of "red water" at the tap . Release of copper from copper pipes can result in customer reports of "blue water" and/or a metallic taste. Release of lead can occur from the solder used to join copper pipe together or from brass fixtures. Copper and lead levels at the consumer's tap are regulated to protect consumer health.

Utilities will often adjust the chemistry of the water before distribution to minimize its corrosiveness. The simplest adjustment involves control of pH and alkalinity to produce a water that tends to passivate corrosion by depositing a layer of calcium carbonate. Corrosion inhibitors are often added to reduce release of metals into the water. Common corrosion inhibitors added to the water are phosphates and silicates.

Maintenance of a biologically safe drinking water is another goal in water distribution. Typically, a chlorine based disinfectant, such as sodium hypochlorite or monochloramine is added to the water as it leaves the treatment plant. Booster stations can be placed within the distribution system to ensure that all areas of the distribution system have adequate sustained levels of disinfection.

Topologies of Water Distribution Networks

Like electric power lines, roads, and microwave radio networks, water systems may have a loop or branch network topology, or a combination of both. The piping networks are circular or rectangular. If any one section of water distribution main fails or needs repair, that section can be isolated without disrupting all users on the network.

Most systems are divided into zones. Factors determining the extent or size of a zone can include hydraulics, telemetry systems, history, and population density. Sometimes systems are designed for a specific area then are modified to accommodate development. Terrain affects hydraulics and some forms of telemetry. While each zone may operate as a stand-alone system, there is usually some arrangement to interconnect zones in order to manage equipment failures or system failures.

Water Network Maintenance

Water supply networks usually represent the majority of assets of a water utility. Systematic documentation of maintenance works using a computerized maintenance management system (CMMS) is a key to a successful operation of a water utility.

Clean drinking water is essential to human life.

Sustainable Urban Water Supply

A sustainable urban water supply network covers all the activities related to provision of potable water. Sustainable development is of increasing importance for the water supply to urban areas. Incorporating innovative water technologies into water supply systems improves water supply from sustainable perspectives. The development of innovative water technologies provides flexibility to the water supply system, generating a fundamental and effective means of sustainability based on an integrated real options approach.

Water is an essential natural resource for human existence. It is needed in every industrial and natural process, for example, it is used for oil refining, for liquid-liquid extraction in hydro-metallurgical processes, for cooling, for scrubbing in the iron and the steel industry and for several operations in food processing facilities , etc. It is necessary to adopt a new approach to design urban water supply networks; water shortages are expected in the forthcoming decades and environmental regulations for water utilization and waste-water disposal are increasingly stringent.

To achieve a sustainable water supply network, new sources of water are needed to be developed, and to reduce environmental pollution.

The price of water is increasing, so less water must be wasted and actions must be taken to prevent pipeline leakage. Shutting down the supply service to fix leaks is less and less tolerated by consumers. A sustainable water supply network must monitor the freshwater consumption rate and the waste-water generation rate.

Many of the urban water supply networks in developing countries face problems related to population increase, water scarcity, and environmental pollution.

Population Growth

In the year 1900 just 13% of the global population lived in cities. This percentage has been rising, and in 2005 49% of the global population lived in urban areas. In 2030 it is predicted, that this statistic will rise to 60% . Attempts to expand water supply by governments are costly and often not sufficient. The building of new illegal settlements makes it hard to map, and make connections to, the water supply, and leads to inadequate water management . In 2002, there were 158 million people with inadequate water supply. An increasing number of people live in slums, in inadequate sanitary conditions, and are therefore at risk of disease.

Water Scarcity

Potable water is not well distributed in the world. 1.8 million deaths are attributed to unsafe water supplies every year, according to the WHO . Many people do not have any access, or do not have access to quality and quantity of potable water, though water itself is abundant. Poor people in developing countries can be close to major rivers, or be in high rainfall areas, yet not have access to potable water at all. There are also people living where lack of water creates millions of deaths every year.

Where the water supply system cannot reach the slums, people manage to use hand pumps, to reach the pit wells, rivers, canals, swamps and any other source of water. In most cases the water quality is unfit for human consumption. The principal cause of water scarcity is the growth in demand. Water is taken from remote areas to satisfy the needs of urban areas. Another reason for water scarcity is climate change: precipitation patterns have changed; rivers have decreased their flow; lakes are drying up; and aquifers are being emptied.

Governmental Issues

In developing countries many governments are corrupt and poor and they respond to these problems with frequently changing policies and non clear agreements. Water demand exceeds supply, and household and industrial water supplies are prioritised over other uses, which leads to water stress. Potable water has a price in the market; water often becomes a business for private companies, which earn a profit by putting a higher price on water, which imposes a barrier for lower-income people. The Millennium Development Goals propose the changes required.

In advanced economies, the problems are about optimising existing supply networks. These economies have usually had continuing evolution, which allowed them to construct infrastructure to supply water to people. The European Union has developed a set of rules and policies to overcome expected future problems.

There are many international documents with interesting, but not very specific, ideas and therefore they are not put into practice . Recommendations have been made by the United Nations, such as the Dublin Statement on Water and Sustainable Development.

Optimizing the Water Supply Network

The yield of a system can be measured by either its value or its net benefit. For a water supply system, the true value or the net benefit is a reliable water supply service having adequate quantity and good quality of the product. For example, if the existing water supply of a city needs to be extended to supply a new municipality, the impact of the new branch of the system must be designed to supply the new needs, while maintaining supply to the old system.

Single-Objective Optimization

The design of a system is governed by multiple criteria, one being cost. If the benefit is *fixed*, the least cost design results in maximum benefit. However, the least cost approach normally results in a *minimum capacity* for a water supply network. A minimum cost model usually searches for the least cost solution (in pipe sizes), while satisfying the hydraulic constraints such as: required output pressures, maximum pipe flow rate and pipe flow velocities. The cost is a function of pipe diameters; therefore the optimization problem consists of finding a minimum cost solution by optimising pipe sizes to provide the minimum acceptable capacity.

Multi-Objective Optimization

However, according to the authors of the paper entitled, “Method for optimizing design and rehabilitation of water distribution systems”, “the least capacity is not a desirable solution to a sustainable water supply network in a long term, due to the uncertainty of the future demand” . It is preferable to provide extra pipe capacity to cope with unexpected demand growth and with water outages. The problem changes from a single objective optimization problem (minimizing cost), to a multi-objective optimization problem (minimizing cost and maximizing flow capacity).

Weighted Sum Method

To solve a multi-objective optimization problem, it is necessary to convert the problem into a single objective optimization problem, by using adjustments, such as a weighted sum of objectives, or an ε-constraint method. The weighted sum approach gives a certain weight to the different objectives, and then factors in all these weights to form a single objective function that can be solved by single factor optimization. This method is not entirely satisfactory, because the weights cannot be correctly chosen, so this approach cannot find the optimal solution for all the original objectives.

The Constraint Method

The second approach (the constraint method), chooses one of the objective functions as the single objective, and the other objective functions are treated as constraints with a limited value. However, the optimal solution depends on the pre-defined constraint limits.

Sensitivity Analysis

The multiple objective optimization problems involve computing the tradeoff between the costs and benefits resulting in a set of solutions that can be used for sensitivity analysis and tested in different scenarios. But there is no single optimal solution that will satisfy the global optimality of both objectives. As both objectives are to some extent contradictory, it is not possible to improve one objective without sacrificing the other. It is necessary in some cases use a different approach,(e.g. Pareto Analysis), and choose the best combination.

Operational Constraints

Returning to the cost objective function, it cannot violate any of the operational constraints. Generally this cost is dominated by the energy cost for pumping. "The operational constraints include the standards of customer service, such us: the minimum delivered pressure, in addition to the physical constraints such us the maximum and the minimum water levels in storage tanks to prevent overtopping and emptying respectively."

In order to optimize the operational performance of the water supply network, at the same time as minimizing the energy costs, it is necessary to predict the consequences of different pump and valve settings on the behavior of the network.

Apart from Linear and Non-linear Programming, there are other methods and approaches to design, to manage and operate a water supply network to achieve sustainability—for instance, the adoption of appropriate technology coupled with effective strategies for operation and maintenance. These strategies must include effective management models, technical support to the householders and industries, sustainable financing mechanisms, and development of reliable supply chains. All these measures must ensure the following: system working lifespan; maintenance cycle; continuity of functioning; down time for repairs; water yield and water quality.

Sustainable Development

In an unsustainable system there is insufficient maintenance of the water networks, especially in the major pipe lines in urban areas. The system deteriorates and then needs rehabilitation or renewal.

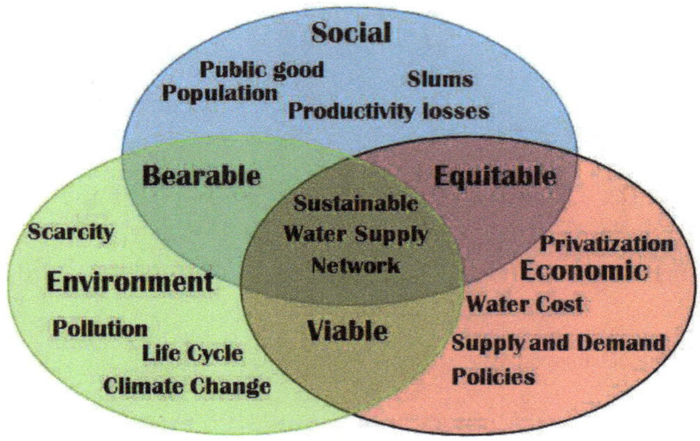

Sustainable development in an urban water network

Householders and sewage treatment plants can both make the water supply networks more efficient and sustainable. Major improvements in eco-efficiency are gained through systematic separation of rainfall and wastewater. Membrane technology can be used for recycling wastewater.

The municipal government can develop a "Municipal Water Reuse System" which is a current approach to manage the rainwater. It applies a water reuse scheme for treated wastewater, on a municipal scale, to provide non-potable water for industry, household and municipal uses. This technology consists in separating the urine fraction of sanitary wastewater, and collecting it for recycling its nutrients. The feces and graywater fraction is collected, together with organic wastes from the households, using a gravity sewer system, continuously flushed with non-potable water. The water is treated anaerobically and the biogas is used for energy production.

The sustainable water supply system is an integrated system including water intake, water utilization, wastewater discharge and treatment and water environmental protection. It requires reducing freshwater and groundwater usage in all sectors of consumption. Developing sustainable water supply systems is a growing trend, because it serves people's long-term interests. . There are several ways to re-use and recycle the water, in order to achieve long-term sustainability, such us:

- Gray water re-use and treatment: gray water is wastewater coming from baths, showers, sinks and washbasins. If this water is treated it can be used as a source of water for uses other than drinking. Depending on the type of gray water and its level of treatment, it can be re-used for irrigation and toilet flushing. According to an investigation about the impacts of domestic grey water reuse on public health, carried out by the New South Wales Health Centre in Australia in the year 2000, grey water contains less nitrogen and fecal pathogenic organisms than sewage, and the organic content of grey water decomposes more rapidly.

- Ecological treatment systems use little energy: there are many applications in gray water re-use, such us reed beds, soil treatment systems and plant filters. This process is ideal for gray water re-use, because of easier maintenance and higher removal rates of organic matter, ammonia, nitrogen and phosphorus.

Other possible approaches to scoping model's for water supply, applicable to any urban area, include the following:

- Sustainable Urban Drainage System.
- Borehole extraction.
- Intercluster groundwater flow.
- Canal and river extraction.
- Aquifer storage
- A more user-friendly indoor water use

The "Dublin Statement on Water and Sustainable Development", mentioned above, is a good example of the new trend to overcome water supply problems. This statement, suggested by advanced economies, has come up with some principles that are of great significance to urban water supply. These are:

- Fresh water is a finite and vulnerable resource, essential to sustain life, development and the environment.
- Water development and management should be based on a participatory approach, involving users, planners and policy-makers at all levels.
- Women play a central part in the provision, management and safeguarding of water. Institutional arrangements should reflect the role of women in water provision and protection
- Water has an economic value in all its competing uses and should be recognized as an economic good. .

From these statements, developed in 1992, several policies have been created to give importance to water and to move urban water system management towards sustainable development. The Water Framework Directive by the European Commission is a good example of what has been created there out of former policies.

Future Approaches

There is great need for a more sustainable water supply systems. To achieve sustainability several factors must be tackled at the same time: climate change, rising energy cost, and rising populations. All of these factors provoke change and put pressure on management of available water resources..

An obstacle to transforming conventional water supply systems, is the amount of time needed to achieve the transformation. More specifically, transformation must be implemented by municipal legislation bodies, which always need short-term solutions too. Another obstacle to achieving sustainability in water supply systems is the insufficient practical experience with the technologies required, and the missing know-how about the organization and the transition process.

Possible ways to improve this situation is simulating of the network, implementing pilot projects, learning from the costs involved and the benefits achieved.

SANITARY SEWER

A sanitary sewer or "foul sewer" is an underground carriage system specifically for transporting sewage from houses and commercial buildings through pipes to treatment or disposal. Sanitary sewers are part of an overall system called sewerage or sewage system.

Sewage may be treated to control water pollution before discharge to surface waters. Sanitary sewers serving industrial areas also carry industrial wastewater.

Separate sanitary sewer systems are designed to transport sewage alone. In municipalities served by sanitary sewers, separate storm drains may be constructed to convey surface runoff directly to surface waters. Sanitary sewers are distinguished from combined sewers, which combine sewage with stormwater runoff in the same pipe. Sanitary sewer systems are considered beneficial because they avoid combined sewer overflows.

PVC sanitary sewer installation. Sanitary sewers are sized to carry the amount of sewage generated by the collection area. Sanitary sewers are much smaller than combined sewers designed to also carry surface runoff.

Background

Sewage treatment is less effective when sanitary waste is diluted with stormwater, and combined sewer overflows occur when runoff from heavy rainfall or snowmelt exceeds hydraulic capacity of sewage treatment plants. To overcome these disadvantages, some cities built separate sanitary sewers to collect only municipal wastewater and exclude stormwater runoff collected in separate storm drains. The decision between a combined sewer system or two separate systems is mainly based on need for sewage treatment and cost of providing treatment during heavy rain events. Many cities with combined sewer systems built prior to sewage treatment have not replaced those sewer systems.

Types

Conventional Gravity Sewers

Manhole access to sewer; person shows scale

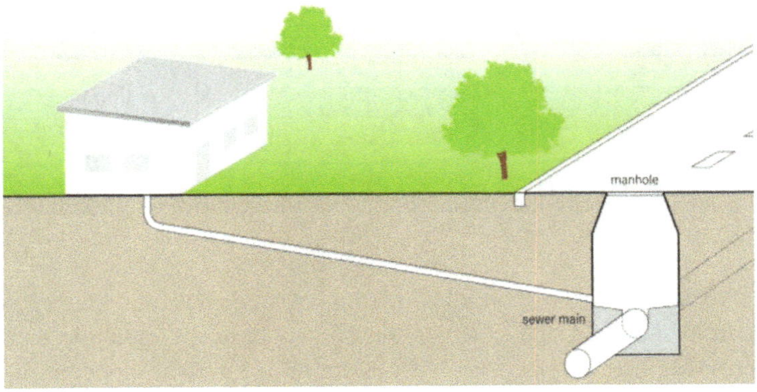

Schematic of a conventional sanitary sewer to convey blackwater and greywater from households to a centralized sewage treatment facility

A manhole cover for a sanitary sewer access point.

View looking down into an open manhole showing two converging sanitary sewer lines. The larger line enters from the right and changes direction within the manhole to exit from the top of the photo. A smaller line enters from the bottom of the photo under the access steps. The concrete floor of the manhole has channels to minimize accumulation of solids.

Interior photo of a large sanitary sewer from an access manhole.

In the developed world, sewers are pipes from buildings to one or more levels of larger underground trunk mains, which transport the sewage to sewage treatment facilities. Vertical pipes, usually made of precast concrete, called manholes, connect the mains to the surface. Depending upon site application and use, these vertical pipes can be cylindrical, eccentric or concentric. The manholes are used for access to the sewer pipes for inspection and maintenance, and as a means to vent sewer gases. They also facilitate vertical and horizontal angles in otherwise straight pipelines.

Pipes conveying sewage from an individual building to a common gravity sewer line are called laterals. Branch sewers typically run under streets receiving laterals from buildings along that street and discharge by gravity into trunk sewers at manholes. Larger cities may have sewers called interceptors receiving flow from multiple trunk sewers.

Design and sizing of sanitary sewers considers the population to be served over the anticipated life of the sewer, per capita wastewater production, and flow peaking from timing of daily routines. Minimum sewer diameters are often specified to prevent blockage by solid materials flushed down toilets; and gradients may be selected to maintain flow velocities generating sufficient turbulence to minimize solids deposition within the sewer. Commercial and industrial flows are also considered, but diversion of surface runoff to storm drains eliminates wet weather flow peaks of inefficient combined sewers.

Force Mains

Pumps may be necessary where gravity sewers serve areas at lower elevations than the sewage treatment plant, or distant areas at similar elevations. A lift station is a gravity sewer sump with a pump to lift accumulated sewage to a higher elevation. The pump may discharge to another gravity sewer at that location or may discharge through a pressurized force main to some distant location.

Effluent Sewer

Effluent sewer systems, also called septic tank effluent drainage (STED) or solids-free sewer (SFS) systems, have septic tanks that collect sewage from residences and businesses, and the effluent that comes out of the tank is sent to either a centralized sewage treatment plant or a distributed treatment system for further treatment. Most of the solids are removed by the septic tanks, so the treatment plant can be much smaller than a typical plant. In addition, because of the vast reduction in solid waste, a pumping system can be used to move the wastewater rather than a gravity system. The pipes have small diameters (typically 1.5" to 4"). Because the wastestream is pressurized, they can be laid just below the ground surface along the land's contour.

Simplified Sewer

Simplified sanitary sewers consist of small-diameter pipes (typically 100 mm or about 4 inches), often laid at fairly flat gradients (1 in 200). The investment cost for simplified sanitary sewers can be about half the cost of conventional sewers. However, the requirements for operation and maintenance are usually higher. Simplified sewers are most common in Brazil and are also used in a number of other developing countries.

Vacuum Sewer

In low-lying communities, wastewater may be conveyed by vacuum sewer. Pipelines range in size from pipes of six inches (150 mm) in diameter to concrete-lined tunnels of up to thirty feet (10 m) in diameter. A low pressure system uses a small grinder pump located at each point of connection, typically a house or business. Vacuum sewer systems use differential atmospheric pressure to move the liquid to a central vacuum station.

Maintenance

Sanitary sewer overflow can occur due to blocked or broken sewer lines, infiltration of excessive stormwater or malfunction of pumps. In these cases untreated sewage is discharged from a sanitary sewer into the environment prior to reaching sewage treatment facilities. To avoid this, maintenance is required.

The maintenance requirements vary with the type of sanitary sewer. In general, all sewers deteriorate with age, but infiltration/inflow is a problem unique to sanitary sewers, since both combined sewers and storm drains are sized to carry these contributions. Holding infiltration to acceptable levels requires a higher standard of maintenance than necessary for structural integrity considerations of combined sewers. A comprehensive construction inspection program is required to prevent inappropriate connection of cellar, yard, and roof drains to sanitary sewers. The probability of inappropriate connections is higher where combined sewers and sanitary sewers are found in close proximity, because construction personnel may not recognize the difference. Many older cities still use combined sewers while adjacent suburbs were built with separate sanitary sewers.

For decades, when sanitary sewer pipes cracked or experienced other damage, the only option was an expensive excavation, removal and replacement of the damaged pipe, typically requiring street repavement afterwards. In the mid-1950s a unit was invented where two units at each end with a special cement mixture in between was pulled from one manhole cover to the next, coating the pipe with the cement under high pressure which then cured rapidly, sealing all cracks and breaks in the pipe. Today, a similar method using epoxy resin is used by some municipalities to re-line aging or damaged pipes, effectively creating a "pipe in a pipe". These methods may be unsuitable for locations where the full diameter of the original pipe is required to carry expected flows, and may be an unwise investment if greater wastewater flows may be anticipated from population growth, increased water use, or new service connections within the expected service life of the repair.

Another popular method for replacing aged or damaged lines is called Pipe bursting, where a new pipe (typically PVC or ABS) is drawn thru the old pipe behind an "expander head" that breaks apart the old pipe as the new one is drawn through behind it.

These methods are most suitable for trunk sewers, since repair of lines with lateral connections is complicated by making provisions to receive lateral flows without accepting undesirable infiltration from inadequately sealed junctions.

Sanitary sewers evolved from combined sewers built where water was plentiful. Animal feces accumulated on city streets while animal-powered transport moved people and goods. Accumulations of animal feces encouraged dumping chamber pots into streets where night soil collection was impractical. Combined sewers were built to use surface runoff to flush waste off streets and move it underground to places distant from populated areas. Sewage treatment became necessary as population expanded, but treatment of diluted waste from combined sewers is more expensive than treating undiluted sewage.

Modern municipal engineering finds its origins in the 19th-century United Kingdom, following the Industrial Revolution and the growth of large industrial cities. The threat to urban populations from epidemics of waterborne diseases such as cholera and typhus lead to the development of a profession devoted to "sanitary science" that later became "municipal engineering".

A key figure of the so-called "public health movement" was Edwin Chadwick, author of the parliamentary report "The Sanitary Condition of the Labouring Population", published in 1842.

Early British legislation included:

- Burgh Police Act 1833 - powers of paving, lighting, cleansing, watching, supplying with water and improving their communities.
- Municipal Corporations Act 1835
- Public Health Act 1866 – formation of drainage boards
- Public Health Act 1875 known at the time as the Great Public Health Act

This legislation provided local authorities with powers to undertake municipal engineering projects and to appoint borough surveyors (later known as "municipal engineers").

In the U.K, the Association of Municipal Engineers, (subsequently named Institution of Municipal Engineers), was established in 1874 under the encouragement of the Institution of Civil Engineers, to address the issue of the application of sanitary science. By the early 20th century, Municipal Engineering had become a broad discipline embracing many of the responsibilities undertaken by local authorities, including roads, drainage, flood control, coastal engineering, public health, waste management, street cleaning, water supply, sewers, waste water treatment, crematoria, public baths, slum clearance, town planning, public housing, energy supply, parks, leisure facilities, libraries, town halls and other municipal buildings.

In the UK, the development of different strands of knowledge necessary for the management of municipal infrastructure led to the emergence of separate specialised institutions, including:

- For drainage: Chartered Institution of Water and Environmental Management, 1895
- For town planning: Town Planning Institute 1914 ... subsequently becoming the Royal Town Planning Institute
- For street lighting: Association of Public Lighting Engineers, 1934...subsequently becoming the Institution of Lighting Engineers
- For highway engineering: Institution of Highways and Transportation, 1930
- For public housing: Institute of Housing, 1931

In 1984 the Institution of Municipal Engineers merged with the Institution of Civil Engineers.

Since the 1970s, there has been a global trend toward increasing privatisation and outsourcing of municipal engineering services.

In the UK in the 1990s a change in management philosophy brought the demise of the traditional organisational structure of boroughs where the three functions of town clerk, borough treasurer and borough engineer were replaced by an administrative structure with a larger number of specialised departments.

In the late 1990s and early 21st century there was increasing dissatisfaction over what was perceived to be fractured and dysfunctional public services designed along narrow specialties. A more holistic approach to urban engineering began to emerge as an alternative concept. Critics of the specialised approach included the Commission for Architecture and the Built Environment that complained that the specialised approach to management of the public realm focussed too much on the on efficient movement of vehicles rather than the more general interests of local communities....GANESH.One of the civil eng.

Professional Practice

In the United Kingdom there is no longer any formal professional qualification in municipal engineering although there are degree courses available in urban engineering.

A professional certificate in Urban Engineering is available from through the Institution of Incorporated Engineers via the Public Realm Information and Advice Network.

The British Institution of Civil Engineers (ICE) caters to practitioners employed in the public sector, private consultancy and academia through its Proceedings Journal Municipal Engineer. The journal, first published in 1873, has a global scope and covers the whole life cycle of municipal services addressing technical, political and community issues. In addition an Expert Panel responds on behalf of ICE to Government consultations and is represented on the International Federation of Municipal Engineering.

International Organization

The International Federation of Municipal Engineering (IFME) is an organisation comprising professional municipal engineers from all round the world. IFME's mission is to connect municipal engineers, public works professionals, public agencies, institutions and businesses around the world in order that they can share a global pool of knowledge and experience. The aim is to foster continued improvement in the quality of public works and wider community services.

The inaugural meeting was held in 1960 at the UNESCO headquarters in Paris. Membership has grown steadily and in 2009 comprised representatives from national associations in: Australia, Canada, Denmark, Estonia, Finland, Italy, Israel, The Netherlands, New Zealand, Norway, Southern Africa (South Africa, Botawana, Namibia & Zimbabwe), Sweden, UK (England, Scotland, Wales & Northern Ireland) and USA. Belgium and San Marino are presently Corresponding Members.

Related Engineering Disciplines

Municipal or urban engineering combines elements of environmental engineering, water resources engineering and transport engineering.

Relationship to Urban Design or Urban Planning

Today, municipal engineering may be confused with urban design or urban planning. Whereas the urbanist or urban planner may design the general layout of streets and public places, the municipal engineer is concerned with the detailed design. For example, in the case of the design of a new street, the urbanist may specify the general layout of the street, including landscaping, surface finishings and urban accessories, but the municipal engineer will prepare the detailed plans and specifications for the roads, sidewalks, municipal services and street lighting.

URBAN DESIGN

Urban design is the process of designing and shaping cities, towns and villages. In contrast to architecture, which focuses on the design of individual buildings, urban design deals with the larger scale of groups of buildings, streets and public spaces, whole neighborhoods and districts, and entire cities, with the goal of making urban areas functional, attractive, and sustainable.

Urban design is an inter-disciplinary subject that utilizes elements of many built environment professions, including landscape architecture, urban planning, architecture, civil and municipal engineering. It is common for professionals in all these disciplines to practice in urban design. In more recent times different sub-strands of urban design have emerged such as strategic urban design, landscape urbanism, water-sensitive urban design, and sustainable urbanism.

Urban design demands an understanding of a wide range of subjects from physical geography, through to social science, and an appreciation for disciplines, such as real estate development, urban economics, political economy and social theory.

Urban design is about making connections between people and places, movement and urban form, nature and the built fabric. Urban design draws together the many strands of place-making, environmental stewardship, social equity and economic viability into the creation of places with distinct beauty and identity. Urban design draws these and other strands together creating a vision for an area and then deploying the resources and skills needed to bring the vision to life.

Urban design theory deals primarily with the design and management of public space (i.e. the 'public environment', 'public realm' or 'public domain'), and the way public places are experienced and used. Public space includes the totality of spaces used freely on a day-to-day basis by the general public, such as streets, plazas, parks and public infrastructure. Some aspects of privately owned spaces, such as building facades or domestic gardens, also contribute to public space and are therefore also considered by urban design theory. Important writers on urban design theory include Christopher Alexander, Peter Calthorpe, Gordon Cullen, Andres Duany, Jane Jacobs, Mitchell Joachim, Jan Gehl, Allan B. Jacobs, Kevin Lynch, Aldo Rossi, Colin Rowe, Robert Venturi, William H. Whyte, Camillo Sitte, Bill Hillier (Space syntax) and Elizabeth Plater-Zyberk.

Urban Design Center Kashiwa-no-ha

Although contemporary professional use of the term 'urban design' dates from the mid-20th century, urban design as such has been practiced throughout history. Ancient examples of carefully planned and designed cities exist in Asia, Africa, Europe and the Americas, and are particularly well-known within Classical Chinese, Roman and Greek cultures (see Hippodamus of Miletus).

European Medieval cities are often, and often erroneously, regarded as exemplars of undesigned or 'organic' city development. There are many examples of considered urban design in the Middle Ages (see, e.g., David Friedman, *Florentine New Towns: Urban Design in the Late Middle Ages*, MIT 1988). In England, many of the towns listed in the 9th century Burghal Hidage were designed on a grid, examples including Southampton, Wareham, Dorset and Wallingford, Oxfordshire, having been rapidly created to provide a defensive network against Danish invaders. 12th century western Europe brought renewed focus on urbanisation as a means of stimulating economic growth and generating revenue. The burgage system dating from that time and its associated burgage plots brought a form of self-organising design to medieval towns. Rectangular grids were used in the Bastides of 13th and 14th century Gascony, and the new towns of England created in the same period.

Throughout history, design of streets and deliberate configuration of public spaces with buildings have reflected contemporaneous social norms or philosophical and religious beliefs (see, e.g., Erwin Panofsky, *Gothic Architecture and Scholasticism*, Meridian Books, 1957). Yet the link between designed urban space and human mind appears to be bidirectional. Indeed, the reverse impact of urban structure upon human behaviour and upon thought is evidenced by both observational study and historical record. There are clear indications of impact through Renaissance urban design on the thought of Johannes Kepler and Galileo Galilei (see, e.g., Abraham Akkerman, "Urban planning in the founding of Cartesian thought," *Philosophy and Geography* 4(1), 2001). Already René Descartes in his *Discourse on the Method* had attested to the impact Renaissance planned new towns had upon his own thought, and much evidence exists that the Renaissance streetscape was also the perceptual stimulus that had led to the development of coordinate geometry (see, e.g., Claudia Lacour Brodsky, *Lines of Thought: Discourse, Architectonics, and the Origins of Modern Philosophy*, Duke 1996).

The beginnings of modern urban design in Europe are associated with the Renaissance but, especially, with the Age of Enlightenment. Spanish colonial cities were often planned, as were some towns settled by other imperial cultures. These sometimes embodied utopian ambitions as well as aims for functionality and good governance, as with James Oglethorpe's plan for Savannah, Georgia. In the Baroque period the design approaches developed in French formal gardens such as Versailles were extended into urban development and redevelopment. In this period, when modern professional specialisations did not exist, urban design was undertaken by people with skills in areas as diverse as sculpture, architecture, garden design, surveying, astronomy, and military engineering. In the 18th and 19th centuries, urban design was perhaps most closely linked with surveyors (engineers) and architects. The increase in urban populations brought with it problems of epidemic disease, the response to which was a focus on public health, the rise in the UK of municipal engineering and the inclusion in British legislation of provisions such as minimum widths of street in relation to heights of buildings in order to ensure adequate light and ventilation.

Much of Frederick Law Olmsted's work was concerned with urban design, and the newly formed profession of landscape architecture also began to play a significant role in the late 19th century.

Modern Urban Design

Modern urban design is a part of the wider discipline of Urban planning. Indeed, Urban planning began as a movement primarily occupied with matters of public health and urban design.

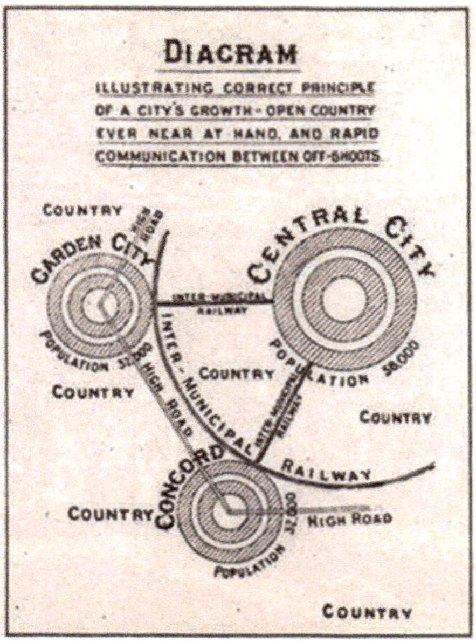

Ebenezer Howard's influential 1902 diagram, illustrating urban growth through garden city "off-shoots"

Planning and architecture went through a paradigm shift at the turn of the 20th century. The industrialized cities of the 19th century had grown at a tremendous rate, with the pace and style of building largely dictated by private business concerns. The evils of urban life for the working poor were becoming increasingly evident as a matter for public concern. The laissez-faire style of government management of the economy, in fashion for most of the Victorian era, was starting to give way to a New Liberalism that championed intervention on the part of the poor and disadvantaged. Around 1900, theorists began developing urban planning models to mitigate the consequences of the industrial age, by providing citizens, especially factory workers, with healthier environments.

The first modern urban planning theorist was Sir Ebenezer Howard, who initiated the garden city movement in 1898. Howard's ideas, although utopian were also highly practical and were adopted around the world in the ensuing decades. His garden cities were intended to be planned, self-contained communities surrounded by parks, containing proportionate and separate areas of residences, industry and agriculture. Inspired by the Utopian novel *Looking Backward* and Henry George's work *Progress and Poverty*, Howard published his book *Garden Cities of To-morrow* in 1898, commonly regarded as the most important book in the history of urban planning. His idealised garden city would house 32,000 people on a site of 6,000 acres (2,428 ha), planned on a concentric pattern with open spaces, public parks and six radial boulevards, 120 ft (37 m) wide, extending from the centre. The garden city would be self-sufficient and when it reached full population, another garden city would be developed nearby. Howard envisaged a cluster of several garden cities as satellites of a central city of 50,000 people, linked by road and rail.

The first garden cities were created at Letchworth and Welwyn Garden City in Hertfordshire. Sir Frederic Osborn extended the movement to regional planning.

Urban planning became professionalized at this period, with input from utopian visionaries as well as from the practical minded infrastructure engineers and local councilors combining to produce new design templates for political consideration. The Town and Country Planning Association was founded in 1899 and the first academic course on urban planning was offered by the University of Liverpool in 1909.

The first official consideration of these new trends was embodied in the Housing and Town Planning Act of 1909 that compelled local authorities to introduce coherent systems of town planning across the country using the new principles of the 'garden city', and to ensure that all housing construction conformed to specific building standards.

Following this Act, surveyors, civil engineers, architects, lawyers and others began working together within local government in the United Kingdom to draw up schemes for the development of land and the idea of town planning as a new and distinctive area of expertise began to be formed. In 1910, Thomas Adams was appointed as the first Town Planning Inspector at the Local Government Board, and began meeting with practitioners. The Town Planning Institute was established in 1914 with a mandate to advance the study of town-planning and civic design. The first university course in America was established at Harvard University in 1924.

The automobile was an important influence on the design of urban development in the 20th century, and the rise of the "urban design" movement can be seen in part as a reaction to the adverse impact of car-use and car orientated design. 'Urban design' was first used as a distinctive term when Harvard University hosted a series of Urban Design Conferences from 1956 . These conferences provided a platform for the launching of Harvard's Urban Design program in 1959-60. The writings of Jane Jacobs, Kevin Lynch, Gordon Cullen and Christopher Alexander became authoritative works for the school of Urban Design.

Gordon Cullen's *The Concise Townscape*, first published in 1961, also had a great influence on many urban designers. Cullen examined the traditional artistic approach to city design of theorists such as Camillo Sitte, Barry Parker and Raymond Unwin. He created the concept of 'serial vision', defining the urban landscape as a series of related spaces.

Jane Jacobs' *The Death and Life of Great American Cities*, published in 1961, was also a catalyst for interest in ideas of urban design. She critiqued the Modernism of CIAM, and asserted that the publicly unowned spaces created by the 'city in the park' notion of Modernists was one of the main reasons for the rising crime rate. She argued instead for an 'eyes on the street' approach to town planning, and the resurrection of main public space precedents, such as streets and squares, in the design of cities.

Kevin Lynch's *The Image of the City* of 1961 was also seminal to the movement, particularly with regards to the concept of legibility, and the reduction of urban design theory to five basic elements - paths, districts, edges, nodes, landmarks. He also made popular the use of mental maps to understanding the city, rather than the two-dimensional physical master plans of the previous 50 years.

Other notable works include Rossi's *Architecture of the City* (1966), Venturi's *Learning from Las Vegas* (1972), Colin Rowe's *Collage City* (1978), and Peter Calthorpe's *The Next American Metropolis* (1993). Rossi introduced the concepts of 'historicism' and 'collective memory' to urban design, and proposed a 'collage metaphor' to understand the collage of new and older forms within the same urban space. Calthorpe, on the other hand, developed a manifesto for sustainable urban living via medium density living, as well as a design manual for building new settlements in accordance with his concept of Transit Oriented Development (TOD). Bill Hillier and Julienne Hanson in "The Social Logic of Space" (1984) introduced the concept of Space Syntax to predict how movement patterns in cities would contribute to urban vitality, anti-social behaviour and economic success. The popularity of these works resulted in terms such as 'historicism', 'sustainability', 'livability', 'high quality of urban components', etc. become everyday language in the field of urban planning.

Current Trends

Jakriborg in Sweden, started in the late 1990s as a new urbanist eco-friendly new town near Malmö

Various current movements in urban design seek to create sustainable urban environments with long-lasting structures, buildings and a great liveability for its inhabitants. The most clearly defined form of walkable urbanism is known as the *Charter of New Urbanism*. It is an approach for successfully reducing environmental impacts by altering the built environment to create and preserve smart cities which support sustainable transport. Residents in compact urban neighborhoods drive fewer miles, and have significantly lower environmental impacts across a range of measures, compared with those living in sprawling suburbs. The concept of Circular flow land use management has also been introduced in Europe to promote sustainable land use patterns, that strive for compact cities and a reduction of greenfield land taken by urban sprawl.

In sustainable construction the recent movement of New Classical Architecture promotes a sustainable approach towards urban construction that appreciates and develops smart growth, walkability, architectural tradition and classical design.This in contrast to modernist and globally uniform architecture, as well as opposing solitary housing estates and suburban sprawl. Both trends started in the 1980s.

Principles

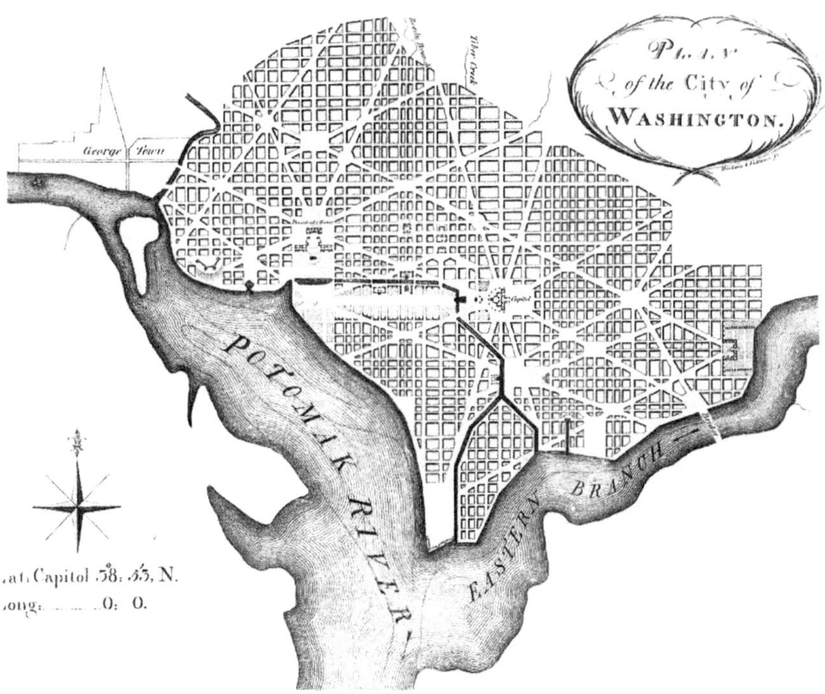

L'Enfant's plan for Washington DC

Gehl Architects' project for Brighton New Road employing shared space

Public spaces are frequently subject to overlapping management responsibilities of multiple public agencies or authorities and the interests of nearby property owners, as well as the requirements of multiple and sometimes competing users. The design, construction and management of public spaces therefore typically demands consultation and negotiation across a variety of spheres. Urban designers rarely have the degree of artistic liberty or control sometimes offered in design professions such as architecture. It also typically requires interdisciplinary input with balanced representation of multiple fields including engineering, ecology, local history, and transport planning.

The scale and degree of detail considered varies depending on context and needs. It ranges from the layout of entire city regions, cities, as with l'Enfant's plan for Washington DC, Griffin and Mahony's plan for Canberra and Doxiadis' plan for Islamabad (although such opportunities are obviously rare), through 'managing the sense of a region' as described by Kevin Lynch, to the design of street furniture.

Urban design may encompass the preparation of design guidelines and regulatory frameworks, or even legislation to control development, advertising, etc. and in this sense overlaps with urban planning. It may encompass the design of particular spaces and structures and in this sense overlaps with architecture, landscape architecture, highway engineering and industrial design. It may also deal with 'place management' to guide and assist the use and maintenance of urban areas and public spaces.

Most urban design work is undertaken by urban planners, landscape architects and architects. There are also professionals who identify themselves specifically as urban designers. Many architecture, landscape and planning programs incorporate urban design theory and design subjects into their curricula and there are an increasing number of university programs offering degrees in urban design, usually at post-graduate level.

Urban design considers:

- Pedestrian zones
- Incorporation of nature within a city
- Aesthetics
- *Urban structure* – How a place is put together and how its parts relate to each other
- *Urban typology, density* and sustainability - spatial types and morphologies related to intensity of use, consumption of resources and production and maintenance of viable communities
- *Accessibility* – Providing for ease, safety and choice when moving to and through places
- *Legibility and wayfinding* – Helping people to find their way around and understand how a place works
- *Animation* – Designing places to stimulate public activity
- *Function and fit* – Shaping places to support their varied intended uses
- *Complementary mixed uses* – Locating activities to allow constructive interaction between them
- *Character and meaning* – Recognizing and valuing the differences between one place and another
- *Order and incident* – Balancing consistency and variety in the urban environment in the interests of appreciating both
- *Continuity and change* – Locating people in time and place, including respect for heritage and support for contemporary culture
- *Civil society* – Making places where people are free to encounter each other as civic equals, an important component in building social capital

Equality Issues

Until the 1970s, the design of towns and cities took little account of the needs of people with disabilities. At that time, disabled people began to form movements demanding recognition of their potential contribution if social obstacles were removed. Disabled people challenged the 'medical model' of disability which saw physical and mental problems as an individual 'tragedy' and people with disabilities as 'brave' for enduring them. They proposed instead a 'social model' which said that barriers to disabled people result from the design of the built environment and attitudes of able-bodied people. 'Access Groups' were established composed of people with disabilities who audited their local areas, checked planning applications and made representations for improvements. The new profession of 'access officer' was established around that time to produce guidelines based on the recommendations of access groups and to oversee adaptations to existing buildings as well as to check on the accessibility of new proposals. Many local authorities now employ access officers who are regulated by the Access Association. A new chapter of the Building Regulations (Part M) was introduced in 1992. Although it was beneficial to have legislation on this issue the requirements were fairly minimal but continue to be improved with ongoing amendments. The Disability Discrimination Act 1995 continues to raise awareness and enforce action on disability issues in the urban environment.

URBAN PLANNING

Urban planning is a technical and political process concerned with the development and use of land, protection and use of the environment, public welfare, and the design of the urban environment, including air, water, and the infrastructure passing into and out of urban areas, such as transportation, communications, and distribution networks. Urban planning is also referred to as urban and regional planning, regional planning, town planning, city planning, rural planning or some combination in various areas worldwide. It takes many forms and it can share perspectives and practices with urban design.

Urban planning guides orderly development in urban, suburban and rural areas. Although predominantly concerned with the planning of settlements and communities, urban planning is also responsible for the planning and development of water use and resources, rural and agricultural land, parks and conserving areas of natural environmental significance. Practitioners of urban planning are concerned with research and analysis, strategic thinking, architecture, urban design, public consultation, policy recommendations, implementation and management.

Urban planners work with the cognate fields of architecture, landscape architecture, civil engineering, and public administration to achieve strategic, policy and sustainability goals. Early urban planners were often members of these cognate fields. Today urban planning is a separate, independent professional discipline. The discipline is the broader category that includes different sub-fields such as land-use planning, zoning, economic development, environmental planning, and transportation planning.

Urban planning designs settlements, from the smallest towns to the largest cities. Shown here is Hong Kong from Western District overlooking Kowloon, across Victoria Harbour.

There is evidence of urban planning and designed communities dating back to the Mesopotamian, Indus Valley, Minoan, and Egyptian civilizations in the third millennium BCE. Archeologists studying the ruins of cities in these areas find paved streets that were laid out at right angles in a grid pattern. The idea of a planned out urban area evolved as different civilizations adopted it. Beginning in the 8th century BCE, Greek city states were primarily centered on orthogonal (or grid-like) plans.The ancient Romans, inspired by the Greeks, also used orthogonal plans for their cities. City planning in the Roman world was developed for military defense and public convenience. The spread of the Roman Empire subsequently spread the ideas of urban planning. As the Roman Empire declined, these ideas slowly disappeared. However, many cities in Europe still held onto the planned Roman city center. Cities in Europe from the 9th to 14th centuries, often grew organically and sometimes chaotically. But many hundreds of new towns were newly built according to preconceived plans, and many others were enlarged with newly planned extensions. Most of these were realized from the 12th to 14th centuries, with a peak-period at the end of the 13th. From the 15th century on, much more is recorded of urban design and the people that were involved. In this period, theoretical treatises on architecture and urban planning start to appear in which theoretical questions are addressed and designs of towns and cities are described and depicted. During the Enlightenment period, several European rulers ambitiously attempted to redesign capital cities. During the Second French Republic, Baron Georges-Eugène Haussmann, under the direction of Napoleon III, redesigned the city of Paris into a more modern capital, with long, straight, wide boulevards.

Planning and architecture went through a paradigm shift at the turn of the 20th century. The industrialized cities of the 19th century grew at a tremendous rate. The pace and style of this industrial construction was largely dictated by the concerns of private business. The evils of urban life for the working poor were becoming increasingly evident as a matter for public concern. The laissez-faire style of government management of the economy, in fashion for most of the Victorian era, was starting to give way to a New Liberalism that championed intervention on the part of the poor and disadvantaged. Around 1900, theorists began developing urban planning models to mitigate the consequences of the industrial age, by providing citizens, especially factory workers, with healthier environments.

Urban planning started to become professionalized during this time. The Town and Country Planning Association was founded in 1899 and the first academic course on urban planning was offered by the University of Liverpool in 1909. In the 1920s, the ideas of modernism and uniformity began to surface in urban planning, and lasted until the 1970s. Many planners started to believe that the ideas of modernism in urban planning led to higher crime rates and social problems. Urban planners now focus more on individualism and diversity in urban centers.

Theories

Planning theory is the body of scientific concepts, definitions, behavioral relationships, and assumptions that define the body of knowledge of urban planning. There are eight procedural theories of planning that remain the principal theories of planning procedure today: the rational-comprehensive approach, the incremental approach, the transactive approach, the communicative approach, the advocacy approach, the equity approach, the radical approach, and the humanist or phenomenological approach.

Technical Aspects

Technical aspects of urban planning involve the applying scientific, technical processes, considerations and features that are involved in planning for land use, urban design, natural resources, transportation, and infrastructure. Urban planning includes techniques such as: predicting population growth, zoning, geographic mapping and analysis, analyzing park space, surveying the water supply, identifying transportation patterns, recognizing food supply demands, allocating healthcare and social services, and analyzing the impact of land use.

Urban Planners

An urban planner is a professional who works in the field of urban planning for the purpose of optimizing the effectiveness of a community's land use and infrastructure. They formulate plans for the development and management of urban and suburban areas, typically analyzing land use compatibility as well as economic, environmental and social trends. In developing the plan for a community (whether commercial, residential, agricultural, natural or recreational), urban planners must also consider a wide array of issues such as sustainability, air pollution, traffic congestion, crime, land values, legislation and zoning codes.

The importance of the urban planner is increasing throughout the 21st century, as modern society begins to face issues of increased population growth, climate change and unsustainable development. An urban planner could be considered a green collar professional.

REFERENCES

1. Jenkinson, Ian. "Municipal Engineer – the silver anniversary". Proceedings of the Institution of Civil Engineers, Municipal Engineer, Vol 162, ME2, June 2009, pp65-68
2. Buchan, Neil. "Briefing Note - International Federation of Municipal Engineering". Proceedings of the Institution of Civil Engineers, Municipal Engineer, Vol 163, ME3, Sept 2009
3. Boeing; et al. (2014). "LEED-ND and Livability Revisited". Berkeley Planning Journal. 27: 31–55. Retrieved 2015-04-15.
4. Van Assche, K., Beunen, R., Duineveld, M., & de Jong, H. (2013). Co-evolutions of planning and design: Risks and benefits of design perspectives in planning systems. Planning Theory, 12(2), 177-198.
5. "Frederick Law Olmsted". fredericklawolmsted.com.

6. Peter Hall, Mark Tewdwr-Jones (2010). Urban and Regional Planning. Routledge.
7. "To-morrow: A Peaceful Path to Real Reform".
8. Goodall, B (1987), Dictionary of Human Geography, London: Penguin.
9. Hardy 1999, p. 4.
10. History 1899–1999 (PDF), TCPA.
11. "urban planning".
12. "The birth of town planning".
13. RTPI.org.uk, About the RTPI. Retrieved 17 January 2013
14. Ewing, R "Growing Cooler - the Evidence on Urban Development and Climate Change". Retrieved on: 2009-03-16.
15. Charter of the New Urbanism
16. "Beauty, Humanism, Continuity between Past and Future". Traditional Architecture Group. Retrieved 23 March 2014.
17. Issue Brief: Smart-Growth: Building Livable Communities. American Institute of Architects. Retrieved on 2014-03-23.
18. "What is Urban Planning". Retrieved April 24, 2015.
19. Van Assche, K., Beunen, R., Duineveld, M., & de Jong, H. (2013). Co-evolutions of planning and design: Risks and benefits of design perspectives in planning systems. Planning Theory, 12(2), 177-198.
20. Taylor, Nigel (2007). Urban Planning Theory since 1945, London, Sage.
21. "What Is Planning?". www.planning.org. Retrieved 2015-09-28.
22. Davreu, Robert (1978). "Cities of Mystery: The Lost Empire of the Indus Valley". The World's Last Mysteries. (second edition). Sydney: Readers' Digest. pp. 121-129. ISBN 0-909486-61-1.
23. Kolb, Frank (1984). Die Stadt im Altertum. München: Verlag C.H. Beck. pp. 51-141: Morris, A.E.J. (1972). History of Urban Form. Prehistory to the Renaissance. London. pp. 22-23.
24. Boerefijn, Wim (2010). The foundation, planning and building of new towns in the 13th and 14th centuries in Europe. An architectural-historical research into urban form and its creation. Phd. thesis Universiteit van Amsterdam. ISBN 978-90-9025157-8.
25. Jordan, David (1992). "Baron Haussmann and Modern Paris". American Scholar. 61 (1): 99.
26. "urban planning".
27. Smith Morris et al. British Town Planning and Urban Design, 1997, ISBN 0-582-23496-4, Longman, Singapore.
28. "How Planners Use Planning Theory". Retrieved April 24, 2015.

CHAPTER 8

Coastal Structures Design and Management

Chapter Contents

Coastal engineering approaches structural and environmental engineering issues along coastlines, where constant wave action creates a dynamic setting. It blends an understanding of oceanography and the processes behind shoreline formation with engineering, the study of building and maintaining structures in a variety of environments. People in this field can work on activities like securing ports, rehabilitating shorelines, and constructing buildings that will be able to withstand harsh conditions near the shore. They may work for government agencies, educational institutions, nonprofits, and private companies.

INTRODUCTION

Coastal engineering is the branch of civil engineering concerning the specific demands posed by constructing at or near the coast, as well as the development of the coast itself.

The hydrodynamic impact of especially waves, tides, storm surges and tsunamis and (often) the harsh environment of salt seawater are typical challenges for the coastal engineer – as are the morphodynamic changes of the coastal topography, caused both by the autonomous development of the system and man-made changes. The areas of interest in coastal engineering include the coasts of the oceans, seas, marginal seas, estuaries and big lakes.

Besides the design, building and maintenance of **coastal structures**, coastal engineers are often interdisciplinary involved in integrated coastal zone management, also because of their specific knowledge of the hydro- and morphodynamics of the coastal system. This may include providing input and technology for e.g. environmental impact assessment, port development, strategies for coastal defense, land reclamation, offshore wind farms and other energy-production facilities, etc.

Wave attack on Ilfracombe's sea walls during a storm.

INTEGRATED COASTAL ZONE MANAGEMENT

Integrated coastal zone management (ICZM) or Integrated coastal management (ICM) is a process for the management of the coast using an integrated approach, regarding all aspects of the coastal zone, including geographical and political boundaries, in an attempt to achieve sustainability

This concept was born in 1992 during the Earth Summit of Rio de Janeiro. The specifics regarding ICZM is set out in the proceedings of the summit within Agenda 21, Chapter 17.

The European Commission defines the ICZM as follows:-

ICZM is a dynamic, multidisciplinary and iterative process to promote sustainable management of coastal zones. It covers the full cycle of information collection, planning (in its broadest sense), decision making, management and monitoring of implementation. ICZM uses the informed participation and cooperation of all stakeholders to assess the societal goals in a given coastal area, and to take actions towards meeting these objectives.

ICZM seeks, over the long-term, to balance environmental, economic, social, cultural and recreational objectives, all within the limits set by natural dynamics. 'Integrated' in ICZM refers to the integration of objectives and also to the integration of the many instruments needed to meet these objectives. It means integration of all relevant policy areas, sectors, and levels of administration. It means integration of the terrestrial and marine components of the target territory, in both time and space.

To further understand the idea of ICZM several aspects can be defined and further explained. The coastal zone, the concept of sustainability and the term integration all within a coastal management context can be individually defined, while the expectations and framework of ICZM can be further explained. This entry uses the example of the New Zealand national framework to illustrate ICZM.

Defining the Coastal Zone

Defining the Coastal zone is of particular importance to the idea of ICZM. But the fuzziness of borders due to the dynamic nature of the coast makes it difficult to clearly define. Most simply the coast can be thought of as an area of interaction between the land and the ocean. Ketchum (1972) defined the area as:

The band of dry land and adjacent ocean space (water and submerged land) in which terrestrial processes and land uses directly affect oceanic processes and uses, and vice versa.

Issues arise with the diversity of features present on the coast and the spatial scales of the interacting systems. Coasts being dynamic in nature are influenced differently all around the world. Influences such as river systems, may reach far inland increasing the complexity and scale of the zone. These issues make it difficult to clearly identify hinterlands and subscribe any subsequent management.

Whilst acknowledging a physical coastal zone, the inclusion of ecosystems, resources and human activity within the zone is important. It is the human activities that warrant management. These activities are responsible for disrupting the natural coastal systems. To add to the complexity of this zone, administrative boundaries use arbitrary lines that dissect the zone, often leading to fragmented management. This sectored approach focuses on specific activities such as land use and fisheries, often leading to adverse effects in another sector.

The Importance of the Coastal Zone and the Need for Management

The dynamic processes that occur within the coastal zones produce diverse and productive ecosystems which have been of great importance historically for human populations. Coastal margins equate to only 8% of the worlds surface area but provide 25% of global productivity. Stress on this environment comes with approximately 70% of the world's population being within a day's walk of the coast. Two-thirds of the world's cities occur on the coast.

Valuable resources such as fish and minerals are considered to be common property and are in high demand for coastal dwellers for subsistence use, recreation and economic development. Through the perception of common property, these resources have been subjected to intensive and specific exploitation. For example; 90% of the world's fish harvest comes from within national exclusive economic zones, most of which are within the sight of shore. This type of practice has led to a problem that has cumulative effects. The addition of other activities adds to the strain placed on this environment. As a whole, human activity in the coastal zone generally degrades the systems by taking unsustainable quantities of resources. The effects are further exacerbated with the input of pollutant wastes. This provides the need for management. Due to the complex nature of human activity in this zone a holistic approach is required to obtain a sustainable outcome.

The Concept of Sustainability

The concept behind the idea of ICZM is sustainability. For ICZM to succeed, it must be sustainable. Sustainability entails a continuous process of decision making, so there is never an end-state just a readjustment of the equilibrium between development and the protection of the environment. The concept of Sustainability or sustainable development came to fruition in the 1987 report of the World Commission on Environment and Development, Our Common Future. It stated sustainable development is "to meet the needs of the present without compromising the ability of future generations to meet their own needs".

Highlighted are three main standpoints which summarise the idea of Sustainable development, they are:

- Economic development to improve the quality of life of people
- Environmentally appropriate development
- Equitable development

To simplify these points, sustainability should acknowledge the right of humans to live a life that is healthy and productive. It should allow for equal distribution of benefits to all people and in doing so protect the environment through appropriate use.

Sustainability is by no means a set of prescriptive actions, more accurately it is a way of thinking. Adapting this way of thinking paves the way for a longer-term view with a more holistic approach, something successful ICZM can achieve.

Expectations of ICZM

As previously stated, for ICZM to be successful it must adhere to the principles that define sustainability and act upon them in ways that are integrated. An optimal balance between environmental protection and the development of economic and social sectors is paramount. As part of the holistic approach ICZM applies, many aspects within a coastal zone are expected to be considered and accounted for. These include but are not limited to: the spatial, functional, legal, policy, knowledge, and participation dimensions. Below are four identified goals of ICZM:

- Maintaining the functional integrity of the coastal resource systems;
- Reducing resource-use conflicts;

- Maintaining the health of the environment;
- Facilitating the progress of multisectoral development

Failure to include these aspects and goals would lead to a form of unsustainable management, undermining the paradigms explicit to ICZM.

Defining Integration

The term 'integration' can be adopted for many different purposes, it is therefore quite important to define the term in the context of the management of the coastal zone to appreciate the intentions of ICZM. Integration within ICZM occurs in and between many different levels, 5 types of integration that occur within ICZM, are explained below;

Integration among sectors: Within the coastal environment there are many sectors that operate. These human activities are largely economic activities such as tourism, fisheries, and port companies. A sense of co-operation between sectors is the main requirement for sector integration within ICZM. This comes from the realisation of a common goal focused around sustainability and the appreciation of one another within the area.

Integration between land and water elements of the coastal zone: This is the realisation of the physical environment being a whole. The coastal environment is a dynamic relationship between many processes all of which are interdependent. The link must be made between imposing a change on one system or feature and its inevitable 'flow on' effects.

Integration among levels of government: Between levels of governance, consistency and co-operation is needed throughout planning and policy making. ICZM is most effective where initiatives have common purpose at local, regional, and national levels. Common goals and actions increase efficiency and mitigate confusion.

Integration between nations: This sees ICZM as an important tool on a global scale. If goals and beliefs are common on a supranational scale, large scale problems could be mitigated or avoided.

Integration among disciplines: Throughout ICZM, knowledge should be accepted from all disciplines. All means of scientific, cultural, traditional, political and local expertise need to be accounted for. By including all these elements a truly holistic approach towards management can be achieved.

The term integration in a coastal management context has many horizontal and vertical aspects, which reflects the complexity of the task and it proves a challenge to implement.

ICZM Framework

Management must embrace a holistic viewpoint of the functions that makeup the complex and dynamic nature of interactions in the coastal environment. Management framework must be applied to a defined geographical limit (often complicated) and should operate with a high level of integration. Due to the diverse nature of the world's coastline and coastal environments, it is not possible to create a framework that is 'one-size-fits-all.'

Different activities, interests and issues also complicate matters. So management will always be unique to countries, regions and ultimately on a local scale.

A common thought process and decision making framework however, can be fairly uniform as a part of ICZM around the world. To achieve the principles set out in sustainable types of management a step by step process can be adhered to.

Firstly, issues and problems need to be identified and assessments of these need to be quantified. This first step will include integration between government, sectoral entities and local residents. The assessments also have to be broad in their application. Once the issues and problems have been identified and weighted, an effective management plan can be made. The plan will be specific to the area in question. Thirdly, the adoption of the plan can be carried out. They can be legally binding statutory plans, strategies or objectives which are generally quite powerful or they can be non-statutory processes and can act as a guide for future development. This duality is largely beneficial as the future can be taken into account, but still provide for a firm stance based in the present. The fourth step is implementation, this active phase includes; law enforcement, education, development etc. The implementation activities will be of course, be as unique as their environments and can take many forms. The last phase is evaluation of the whole process. The principles of sustainability mean that there is no 'end state.' ICZM is an ongoing process which should constantly readjust the equilibrium between economic development and the protection of the environment. Feedback is a crucial part of the process and allows for continued effectiveness even when a situation may change. Public participation and stakeholder involvement is essential in ICZM processes, not only in terms of a democratic approach, but also from a technical–instrumental point of view, in order to reduce decisional conflicts (Ioppolo et al., 2013).

ICZM in the Mediterranean

At the Conference of the Plenipotentiaries on the ICZM Protocol that took place on 20–21 January 2008 in Madrid, the ICZM Protocol was signed. Under the presidency of the Minister of Environment of Spain, H.E. Ms. Cristina Narbona Ruiz, fourteen Contracting Parties of the Barcelona Convention signed the Protocol. These are the following: Algeria, Croatia, France, Greece, Israel, Italy, Malta, Monaco, Montenegro, Morocco, Slovenia, Spain, Syria and Tunisia. All other Parties announced to do so in the very near future. This is the 7th Protocol in the framework of the Barcelona Convention, and the decision to approve the draft text and recommendation to the Conference of the Plenipotentiaries to sign it was taken at the 15th Ordinary Meeting of the Contracting Parties during their meeting in Almeria, on 15–18 January 2008. All the parties are convinced that this Protocol is a crucial milestone in the history of the Mediterranean Action Plan of the United Nations Environment Programme (UNEP/MAP), the first-ever Regional Seas Programme under UNEP's umbrella. It will allow the countries to better manage their coastal zones, as well as to deal with the emerging coastal environmental challenges, such as the climate change.

The ICZM Protocol is a unique legal instrument in the entire international community and the Mediterranean countries are proud of this fact. They are willing to share these experiences with other coastal countries of the world. The signing of the Protocol came after six years of dedicated work of all the Parties. Syria entered history for being the sixth, and "enter-into-force", country for the ICZM Protocol! Namely, the President of the Syrian Arab Republic issued a Legislative Decree No. 85 dated 31 September 2010 for the ratification of the ICZM Protocol. With this 6th ratification, the ICZM Protocol entered into force one month later provided that Syria deposits the instrument of ratification to the depositary country, i.e. Spain. In September 2012, Croatia and Morocco ratified the Protocol, which brought the number of ratifications to 9 (Slovenia, Montenegro, Albania, Spain, France, European Union, Syria, Croatia, Morocco).

The Action Plan for the implementation of the ICZM Protocol 2012-2019 was adopted on the occasion of the CoP 17, held in Paris from 8 to 10 February 2012. The core purposes and objectives of this Action Plan are to implement the Protocol based on country-based planning and regional co-ordination, namely: 1. Support the effective implementation of the ICZM Protocol at regional, national and local levels including through a Common Regional Framework for ICZM; 2. Strengthen the capacities of Contracting Parties to implement the Protocol and use in an effective manner ICZM policies, instruments, tools and processes; and 3. Promote the ICZM Protocol and its implementation within the region, and promote it globally by developing synergies with relevant Conventions and Agreements.

A road-map for the implementation of the ICZM Process, prepared by Priority Actions Programme Regional Activity Centre (PAP/RAC), is available on the Coastal Wiki platform of the PEGASO and ENCORA projects: ICZM Process.

On May 8, 2014 the Israeli Government ratified the ICZM Protocol. This Resolution (#1588) was made in accordance with Article 19(b) of the Government Rules of Procedure. The ICZM Protocol ratification by Israel brings the number of ratifications to 10.

It has become tradition in the Mediterranean to celebrate the Coast Day after it had first been organised in 2007. The main purpose of the Mediterranean Coast Day is to raise awareness of the importance of the coast as both natural and economic resource, as well as to warn of the risks to which it is exposed due to natural phenomena and human actions. September 25 was chosen as the Mediterranean Coast Day to honour the date when Slovenia ratified the ICZM Protocol as the first Mediterranean country to do so. Each year the central Mediterranean celebration is organised in a different country, while other countries organise their own events. Thus, in 2007 and 2008 the central celebration was organised in Sardinia, Italy. In 2009 the host to the central celebration was Turkey, in 2010 Slovenia, in 2011 Algeria, in 2012 Croatia, in 2013 Italy and in 2014 Tunisia. In 2014, a short animated movie "A good climate for change" was produced by PAP/RAC (in co-operation with BONOBO Studio Zagreb, Croatia) for the 2014 Mediterranean Coast Day. The movie was officially launched on the occasion of the central regional celebration of the Mediterranean Coast Day in Tunisia which was organised in the frame of the MedPartnership project.

Constraints of ICZM

Major constraints of ICZM are mostly institutional, rather than technological. The 'top-down' approach of administrative decision making sees problematisation as a tool promoting ICZM through the idea of sustainability. Community-based 'bottom-up' approaches can perceive problems and issues that are specific to a local area. The benefit of this is that the problems are real and acknowledged rather than searched for to fit an imposed strategy or policy. Public consultation and involvement is very important for current 'top-down' approaches, as it can incorporate this 'bottom-up' idea into the policies made. Prescriptive 'top-down' methods have not able to effectively address problems of resource utilization in poor coastal communities as perceptions of the coastal zone differ with regard to developed and developing countries. This leads on to another constraint to ICZM, the idea of common property.

The coastal environment has huge historical and cultural connections with human activity. Its wealth of resources have provided for millennia, with regard to ICZM how does management become legally binding if the dominant perception of the coast is of a common area available to all? And should it? Enforcing restrictions or change to activities within the coastal zone can be difficult as these resources are often very important to people's livelihoods. The idea of the coast being common property fouls 'top-down' approaches. The idea of common property itself is not all that clean, This perception can lead to cumulative exploitation of resources – the very problem this management seeks to extinguish.

ICZM: The New Zealand case study

New Zealand is quite unique as it uses sustainable management within legislation, with a high level of importance placed on to the coastal environment. The Resource Management Act (RMA) (1991) promoted sustainable development and mandated the preparation of a New Zealand Coastal Policy Statement (NZCPS), a national framework for coastal planning. It is the only national policy statement that was mandatory. All subsequent planning must not be inconsistent with the NZCPS, making it a very important document. Regional authorities are required to produce Regional coastal policy plans under the RMA (1991) but strangely enough, they only need to include the marine environment seaward of the mean high water mark. But many regional councils have chosen to integrate the 'dry' landward area within their plans, breaking down the artificial barriers. This attempt at ICZM is still in its early days running into many legislative hurdles and is yet to achieve a fully ecosystems-based approach. But as part of ICZM, evaluation and adoption of changes is important and ongoing changes to the NZCPS in the form of reviews is currently happening. This will provide an excellent stepping stone for future initiatives and the development of a fully integrated form of coastal management.

ICZM: The Iran Case Study

Preparation of comprehensive management plans for optimum utilization of existent sources and potentials in all developed and developing countries is one of the appropriate approaches for constant and permanent utilization of natural, human and financial sources. The versatility of natural sources in coastal areas has made private and governmental users and investors to participate in this section to gain the utmost profits. Therefore, the necessity of preparation and implementation of management plans for perpetual utilization of existent sources in coastal areas has become inevitable. Iran, possessing some 6000 km of coastline in north and south, owns abundant economic capacities in coastal zones and regarding the versatility of nature and coast operators and management of coastal activities and operations, necessity of attention to Integrated Coastal Zone Management becomes more significant. Such necessity has gained its legal support through ratification of arrangements no. 40 from transportation chapter of third and article no. 63 of fourth economic, social and cultural economic schedule and its executive regulations. The General Director of coasts and ports engineering of Ports and Maritime Organization was detailed to take the studies of ICZM into consideration. The first phase of these studies began in spring 2003 and was fulfilled in autumn 2006. The outcome of this phase was compilation of following reports accomplished by several national and international skilled consultants: 1- Project Methodology 2- Scrutinized scope of services related to studies 3- Investigation of studies' needs and project preparation and performance 4- Study, definition and determination of Iranian coastal zones boundaries 6- Investigation of International concepts, methods and experiences about Integrated Coastal Zone Management 7- Study and investigation of different features of Integrated Coastal Zone Management in Iran 8- Preparation and designation of geographic database 9- Purchasing and preparing basic data The second phase of studies started up in autumn 2005 and since then this phase has been fully accomplished and presented, In which six competent Iranian consultants with some cooperation of international consultants are responsible for preparing the eleven results of second part of the studies.

The European Union

The European Parliament and the European Council "adopted in 2002 a **Recommendation** on Integrated Coastal Zone Management which defines the principles sound coastal planning and management. These include the need to base planning on sound and shared knowledge, the need to take a long-term and cross-sector perspective, to pro-actively involve stakeholders and the need to take into account both the terrestrial and the marine components of the coastal zone".

Conclusion

The Integrated Coastal Zone Management (ICZM) appears to be a key element for the sustainable development of these zones. However this recent notion may not be adapted to all cases. The natural disasters Sumatra earthquake and the Indian Ocean tsunami have made a lot of impact on the coastal environment and also the stakeholder's perception on mitigation and management of coastal hazards. Successful implementation is still a major challenge to the idea of ICZM .

Specific Elements of Coastal Engineering

Beach nourishment at the Dutch coast.

The coastal environment produces challenges specific for this branch of engineering: waves, storm surges, tides, tsunamis, sea level changes, sea water and the marine ecosystem.

Most often, in coastal engineering projects there is a need for metocean conditions: local wind and wave climate, as well as statistics for and information on other hydrodynamic quantities of interest. Also, bathymetry and morphological changes are of direct interest. In case of studies of sediment transport and morphological changes, relevant properties of the sea bed sediments, water and ecosystem properties are needed.

Long and Short Waves

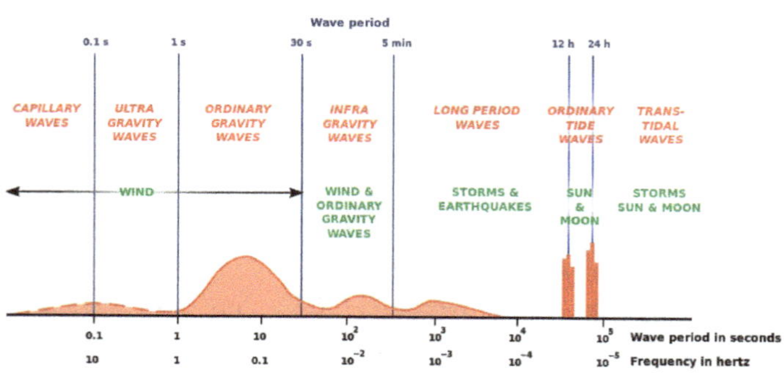

Classification of the spectrum of ocean waves according to wave period, by Walter Munk (1950).

The occurrence of wave phenomena – like sea waves, swell, tides and tsunamis – require engineering knowledge of their physics, as well as models: both numerical models and physical models. The practices in present-day coastal engineering are more-and-more based on models verified and validated by experimental data.

Apart from the wave transformations themselves, for the waves coming from deep water into the shallow coastal waters and surf zone, the effects of the waves are important. These effects include:

- the wave loading on coastal structures like breakwaters, groynes, jetties, sea wallsand dikes
- wave-induced currents, like the longshore current in the surf zone, rip currents and Stokes drift, affecting sediment transport and morphodynamics
- wave agitation in harbors, which may result in harbor downtime
- wave overtopping over seawalls and dikes, which may e.g. threaten the stability of a dike

BREAKWATER (STRUCTURE)

Breakwaters are structures constructed on coasts as part of coastal defense or to protect an anchorage from the effects of both weather and longshore drift.

Breakwaters create safer harbours, but can also trap sediment moving along the coast. Alamitos Bay, California entrance channel.

Purposes of Breakwaters

Breakwaters reduce the intensity of wave action in inshore waters and thereby reduce coastal erosion or provide safe harbourage. Breakwaters may also be small structures designed to protect a gently sloping beach and placed one to three hundred feet offshore in relatively shallow water.

An anchorage is only safe if ships anchored there are protected from the force of high winds and powerful waves by some large underwater barrier which they can shelter behind. Natural harbours are formed by such barriers as headlands or reefs. Artificial harbours can be created with the help of breakwaters. Mobile harbours, such as the D-Day Mulberry harbours, were floated into position and acted as breakwaters. Some natural harbours, such as those in Plymouth Sound, Portland Harbour and Cherbourg, have been enhanced or extended by breakwaters made of rock.

Unintended Consequences

The dissipation of energy and relative calm water created in the lee of the breakwaters often encourage accretion of sediment (as per the design of the breakwater scheme). However, this can lead to excessive salient build up, resulting in tombolo formation, which reduces longshore drift shoreward of the breakwaters. This trapping of sediment can cause adverse effects down-drift of the breakwaters, leading to beach sediment starvation and increased erosion. This may then lead to further engineering protection being needed down-drift of the breakwater development.

Breakwaters are subject to damage, and overtopping in severe storms events.

Three of the four breakwaters forming Portland Harbour

The eight offshore breakwaters at Elmer, UK

Construction

Breakwaters can be constructed with one end linked to the shore, otherwise they are positioned offshore from as little as 100 m up to 300–600 m from the original shoreline. There are two main types of offshore breakwater, single and multiple; single as the name suggests means the breakwater consists of one unbroken barrier, which multiple breakwaters (in numbers anywhere from 2-20) are positioned with gaps in between (50–300 m). Length of gap is largely governed by the interacting wavelengths. Breakwaters may be either fixed or floating, and impermeable or permeable to allow sediment transfer shoreward of the structures, the choice depending on tidal range and water depth. They usually consist of large pieces of rock (granite) weighing up to 16 tonnes each, or rubble-mound. Their design is influenced by the angle of wave approach and other environmental parameters. Breakwater construction can be either parallel or perpendicular to the coast, depending on the shoreline requirements.

Types of Breakwater Structures

A breakwater structure is designed to absorb the energy of the waves that hit it, either by using mass (e.g., with caissons), or by using a revetment slope (e.g., with rock or concrete armour units).

In coastal engineering, a revetment is a land backed structure whilst a breakwater is a sea backed structure (i.e., water on both sides).

Caisson breakwaters typically have vertical sides and are usually erected where it is desirable to berth one or more vessels on the inner face of the breakwater. They use the mass of the caisson and the fill within it to resist the overturning forces applied by waves hitting them. They are relatively expensive to construct in shallow water, but in deeper sites they can offer a significant saving over revetment breakwaters.

Rubble mound breakwaters use structural voids to dissipate the wave energy. Rock or concrete armour units on the outside of the structure absorb most of the energy, while gravels or sands prevent the wave energy's continuing through the breakwater core. The slopes of the revetment are typically between 1:1 and 1:2, depending upon the materials used. In shallow water, revetment breakwaters are usually relatively inexpensive. As water depth increases, the material requirements, and hence costs, increase significantly.

Breakwater Armour Units

As design wave heights get larger, rubble mound breakwaters require larger armour units to resist the wave forces. These armour units can be formed of concrete or natural rock. The largest standard grading for rock armour units given in CIRIA 683 'The Rock Manual' is 10-15 tonnes. Larger gradings may be available, but the ultimate size is limited in practice by the natural fracture properties of locally available rock. Shaped concrete armour units (such as Dolos, Xbloc, Tetrapod, etc.) can be provided in up to approximately 40 tonnes (e.g. Jorf Lasfar, Morocco), before they become vulnerable to damage under self weight, wave impact and thermal cracking of the complex shapes during casting/curing. Where the very largest armour units are required for the most exposed locations in very deep water, armour units are most often formed of concrete cubes, which have been used up to ~195 tonnes for the tip of the breakwater at Punta Langosteira near La Coruña, Spain.

Preliminary design of armour unit size is often undertaken using the Hudson Equation, Van der Meer and more recently Van Gent et al.; these methods are all described in CIRIA 683 'The Rock Manual', the United States Army Corps of Engineers Shore Protection Manual (available for free online) and elsewhere. For detailed design the use of scaled physical hydraulic models remains the most reliable method for predicting real-life behavior of these complex structures.

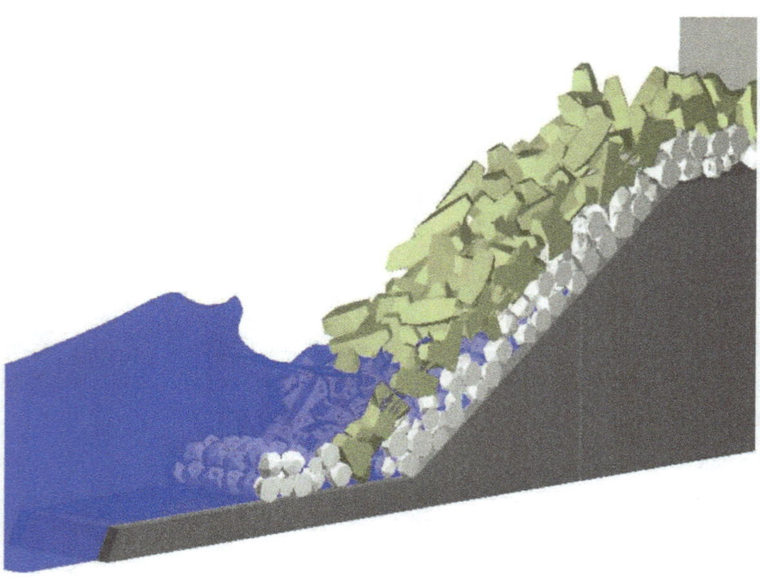

3D simulation of wave motion near a sea wall. MEDUS (2011) Marine Engineering Division of University of Salerno.

The same wave motion at the Visby breakwater in Sweden.

GROYNE

A **groyne** (**groin** in the United States) is a rigid hydraulic structure built from an ocean shore (in coastal engineering) or from a bank (in rivers) that interrupts water flow and limits the movement of sediment. It is usually made out of wood, concrete or stone. In the ocean, groynes create beaches or prevent them being washed away by longshore drift. In a river, groynes prevent erosion and ice-jamming, which in turn aids navigation. Ocean groynes run generally perpendicular to the shore, extending from the upper foreshore or beach into the water. All of a groyne may be under water, in which case it is a **submerged groyne**. The areas between groups of groynes are **groyne fields**. Groynes are generally placed in groups. They are often used in tandem with seawalls. Groynes, however, may cause a shoreline to be perceived as unnatural.

Groynes in Sitges

In Coastal Engineering

Rock groyne

A groyne's length and elevation, and the spacing between groynes is determined according to local wave energy and beach slope. Groynes that are too long or too high tend to accelerate downdrift erosion, and are ineffective because they trap too much sediment. Groynes that are too short, too low, or too permeable are ineffective because they trap too little sediment. If a groyne does not extend far enough landward, water (for example at a high tide combined with a storm surge) may flow past the landward end and erode a channel bypassing the groyne, a process known as *flanking*.

Working

A groyne creates and maintains a wide area of beach or sediment on its updrift side, and reduces erosion on the other. It is a physical barrier to stop sediment transport in the direction of longshore drift (also called longshore transport). This causes a build-up, which is often accompanied by accelerated erosion of the downdrift beach, which receives little or no sand from longshore drift (this is known as terminal groyne syndrome, as it occurs after the **terminal groyne** in a group of groynes). Groynes add sediment to the beach by capturing downward drift. However, this can cause severe erosion on shorelines downstream from the groyne. If a groyne is correctly designed, then the amount of material it can hold will be limited, and excess sediment will be free to move on through the system. However, if a groyne is too large it may trap too much sediment, which can cause severe beach erosion on the down-drift side.

In Rivers

Aerial photograph of the Lumberville Dam on the Delaware River. Some water is pouring over the wing dams or groins; there is a navigation channel between them.

River groynes (**spur dykes**, wing dykes, or wing dams) are often constructed nearly perpendicular to the riverbanks, beginning at a riverbank with a root and ending at the regulation line with a head. They maintain a channel to prevent ice jamming, and more generally improve navigation and control over lateral erosion, that would form from meanders. Groynes have a major impact on the river morphology: they cause autonomous degradation of the river.

They are also used around bridges to prevent bridge scour.

Types

Submerged Groin (US spelling), Hunting Island, South Carolina

A "Keep off the groynes" sign in Brighton with a groyne showing in the background.

Groynes can be distinguished by how they are constructed, whether they are submerged, their effect on stream flow or by shape.

By Construction Method

Groynes can be permeable, allowing the water to flow through at reduced velocities, or impermeable, blocking and deflecting the current.

- **Permeable groynes** are large rocks, bamboo or timber
- **impermeable groynes** (**solid groynes** or **rock armour groynes**) are constructed using rock, gravel, gabions.

By whether they are Submerged

Groynes can be submerged or not under normal conditions. Usually impermeable groynes are non-submerged, since flow over the top of solid groynes may cause severe erosion along the shanks. Submerged groynes, on the other hand, may be permeable depending on the degree of flow disturbance needed.

By Their Effect on Stream Flow

Groynes can be attracting, deflecting or repelling.

- **Attracting groynes** point downstream, serving to attract the stream flow toward themselves and not repel the flow toward the opposite bank. They tend to maintain deep current close to the bank.
- **Deflecting groynes** change the direction of flow without repelling it. They are generally short and used for limited, local protection.
- **Repelling groynes** point upstream; they force the flow away from themselves. A single groyne may have one section, for example, attracting, and another section deflecting.

Groynes can be built with different planview shapes. Some examples include straight groynes, T head, L head, hockey stick, inverted hockey stick groynes, straight groynes with pier head, wing, and tail groynes.

Groynes in the Waal river, part of the Rhine in the Netherlands

Groyne on the East coast of England

Timber groynes in Bournemouth, England

Timber groyne in Sandown, Isle of Wight, England

Groyne in Schobüll, Nordfriesland, Germany

Groyne in Crescent Beach, Canada

Groynes off the Ennore Expressway, near Chennai, India

Groynes on the Vistula river, Warsaw, Poland

Groynes on Arabian Sea at Paravur Pozhikara near Kollam city in India

JETTY

A jetty is a structure that projects from the land out into water. Often, "jetty" refers to a walkway accessing the centre of an enclosed waterbody. The term is derived from the French word jetée, "thrown", and signifies something thrown out.

Coastal lagoons fronted by barrier spits typically have entrances that migrate through time. Here, the entrance has been fixed by jetty construction. Carlsbad, California, April 1998.

For Regulating Rivers

Another form of jetties, wing dams are extended out, opposite one another, *from each bank of a river*, at intervals, to contract a wide channel, and by concentration of the current to produce a deepening.

For Berthing at Docks

Where docks are given sloping sides, openwork timber jetties are generally carried across the slope, at the ends of which vessels can lie in deep water or more solid structures are erected over the slope for supporting coal-tips. Pilework jetties are also constructed in the water outside the entrances to docks on each side, so as to form an enlarging trumpet-shaped channel between the entrance, lock or tidal basin and the approach channel, in order to guide vessels in entering or leaving the docks. Solid jetties, moreover, lined with quay walls, are sometimes carried out into a wide dock,at right angles to the line of quays at the side, to enlarge the accommodation; and they also serve, when extended on a large scale from the coast of a tideless sea under shelter of an outlying breakwater, to form the basins in which vessels lie when discharging and taking in cargoes in such a port as Marseille.

At Entrances to Jetty Harbors

Aerial view of jetties, constructed of dolos, at Humboldt Bay harbor entrance. These jetties mitigate unpredictable shifting sands and ensure 700 feet (210 m) of width.

The approach channel to some ports situated on sandy coasts is guided and protected across the beach by parallel jetties, made solid up to a little above low water of neap tides, on which open timber-work is erected, provided with a planked platform at the top raised above the highest tides. The channel between the jetties was originally maintained by tidal scour from low-lying areas close to the coast, and subsequently by the current from sluicing basins; but it is now often considerably deepened by sand-pump dredging. It is protected to some extent by the solid portion of the jetties from the inroad of sand from the adjacent beach, and from the levelling action of the waves; while the upper open portion serves to indicate the channel and to guide the vessels, if necessary (see harbor). The bottom part of the older jetties, in such long-established jetty ports as Calais, Dunkirk and Ostend, was composed of clay or rubble stone, covered on the top by fascine-work or pitching, but the deepening of the jetty channel by dredging and the need that arose for its enlargement led to the reconstruction of the jetties at these ports. The nes jetties at Dunkirk were founded in the sandy beach, by the aid of compressed air, at a depth of 22.75 feet (6.93 m). below low water of spring tides; and their solid masonry portion, on a concrete foundation was raised 50 feet (15 m). above low water of neap tides.

At Lagoon Outlets

A small tidal rise spreading tidal water over a large expanse of lagoon or inland backwater causes the influx and efflux of the tide to maintain a deep channel through a narrows no longer confined by a bank on each side, becomes dispersed, and owing to the reduction of its scouring force, is no longer able at a moderate distance from the shore effectually to resist the action of tending to form a continuous beach in front of the outlet. Hence a bar is produced that diminishes the available depth in the approach channel. By carrying out a solid jetty over the bar, however on each side of the outlet, the tidal currents are concentrated in the channel across the bar, and lower it by scour. Thus the available depth of the approach channels to Venice through the Malamocco and Lido outlets from the Venetian Lagoon have been deepened several feet (metres) over their bars by jetties of rubble, carried out across the foreshore into deep water on both sides of the channel. Other examples are provided by the long jetties extended into the sea in front of the entrance to Charleston harbour, formerly constructed of fascines weighed down with stone and logs, but subsequently of rubble stone, and by the two converging rubble jetties carried out from each shore of Dublin Bay for deepening the approach to Dublin harbour.

Rivers

North Jetty on the left and South Jetty on the right at the mouth of Sebastian Inlet in Florida from the Indian River to the Atlantic Ocean.

At the Outlet of Tideless Rivers

Jetties have been constructed on each side of the outlet river of some of the rivers flowing into the Baltic, with the objective of prolonging the scour of the river and protecting the channel from being shoaled by the littoral drift along the shore. The most interesting application of parallel jetties is in lowering the bar in front of one of the mouths of a deltaic river flowing into a tide — a virtual prolongation of its less sea, by extending the scour of the river out to the bar by banks. Jetties prolonging the Sulina branch of the Danube into the Black Sea, and the south pass of the Mississippi River into the Gulf of Mexico, formed of rubble stone and concrete blocks, and respectively, have enabled the discharge of these rivers to scour away the bars obstructing the access to them; and they have also carried the sediment-bearing waters sufficiently far out to come under the influence of littoral currents, which, by conveying away some of the sediment, postpone the eventual formation of a fresh bar farther out (see river engineering).

At the Mouth of Tidal Rivers

Where a river is narrow near its mouth, has a generally feeble discharge and a small tidal range, the sea is liable on an exposed coast to block up its outlet during severe storms.

The river is thus forced to seek another exit at a weak spot of the beach, which along a low coast may be at some distance off; and this new outlet in its turn may be blocked up, so that the river from time to time shifts the position of its mouth. This inconvenient cycle of changes may be stopped by fixing the outlet of the river at a suitable site, by carrying a jetty on each side of this outlet across the beach, thereby concentrating its discharge in a definite channel and protecting the mouth from being blocked up by littoral drift. This system was long ago applied to the shifting outlet of the river Yare to the south of Yarmouth, and has also been successfully employed for fixing the wandering mouth of the Adur near Shoreham, and of the Adour flowing into the Bay of Biscay below Bayonne. When a new channel was cut across the Hoek van Holland to provide a straighter and deeper outlet channel for the river Maas, forming the approach channel to Rotterdam, low, broad, parallel jetties, composed of fascine mattresses weighted with stone, were carried across the foreshore into the sea on either side of the new mouth of the river, to protect the jetty channel from littoral drift, and cause the discharge of the river to maintain it out to deep water. The channel, also, beyond the outlet of the river Nervion into the Bay of Biscay has been regulated by jetties; and by extending the southwest jetty out for nearly 0.5 miles (0.80 km) with a curve concave towards the channel the outlet has not only been protected to some extent from the easterly drift, but the bar in front has been lowered by the scour produced by the discharge of the river following the concave bend of the southwest jetty. As the outer portion of this jetty was exposed to westerly storms from the Bay of Biscay before the outer harbour was constructed, it has been given the form and strength of a breakwater situated in shallow water.

SEAWALL

A **seawall** (or **sea wall**) is a form of coastal defense constructed where the sea, and associated coastal processes, impact directly upon the landforms of the coast. The purpose of a sea wall is to protect areas of human habitation, conservation and leisure activities from the action of tides and waves. As a seawall is a static feature it will conflict with the dynamic nature of the coast and impede the exchange of sediment between land and sea.

The coast is generally a high-energy, dynamic environment with spatial variations over a wide range of timescales. The coast is exposed to erosion by rivers and winds as well as the sea, so that a combination of denudational processes will work against a sea wall.Because of these persistent natural forces, sea walls need to be maintained (and eventually replaced) to maintain their effectiveness.

The many types of sea wall in use today reflect both the varying physical forces they are designed to withstand, and location specific aspects, such as local climate, coastal position, wave regime, and value of landform. Sea walls are hard engineering shore-based structures which protect the coast from erosion. But various environmental problems and issues may arise from the construction of a sea wall, including disrupting sediment movement and transport patterns. Combined with a high construction cost, this has led to an increasing use of other soft engineering coastal management options such as beach replenishment.

Sea walls may be constructed from various materials, most commonly reinforced concrete, boulders, steel, or gabions. Other possible construction materials are: vinyl, wood, aluminium, fibreglass composite, and large biodegrable sandbags made of jute and coir. In the UK, *sea wall* also refers to an earthen bank used to create a polder, or a dike construction.

An example of a modern seawall in Ventnor on the Isle of Wight

People socializing and walking at the Malecón, Havana

Seawall at Urangan, Queensland

Types

A seawall works by reflecting incident wave energy back into the sea, thus reducing the energy available to cause erosion. In addition to their ugliness, sea walls have two specific weaknesses. First, wave reflection from the wall may result in scour and subsequent lowering of the sand level of the fronting beach. Second, sea walls may accelerate erosion of adjacent, unprotected coastal areas because they affect the littoral drift process. The design and type of sea wall that is appropriate depends on aspects specific to the location, including the surrounding erosion processes. There are three main types of seawalls: vertical, curved or stepped, and mounds, as set out in the table:

Seawall types

Type	Illustration	Advantages	Disadvantages	Example
Vertical	Vertical seawalls are built in particularly exposed situations. These reflect wave energy. Under storm conditions a non-breaking standing wave pattern can form, resulting in a stationary clapotic wave which moves up and down but does not travel horizontally.[13][14] These waves promote erosion at the toe of the wall and can cause severe damage to the sea wall.[15] In some cases piles are placed in front of the wall to lessen wave energy slightly.			
	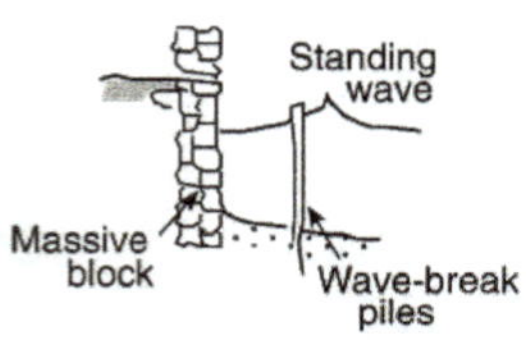	The first implemented, most easily designed and constructed type of seawall. Vertical sea walls deflect wave energy away from the coast. Loose rubble can absorb wave energy.	These can suffer a lot of expensive damage in a short period of time. Vertical design can be undercut by high-wave energy environments over a long period of time.	
Curved	Curved or stepped seawalls are designed to enable waves to break to dissipate wave energy and to repel waves back to the sea. The curve can also prevent the wave overtopping the wall and provides additional protection for the toe of the wall.			
	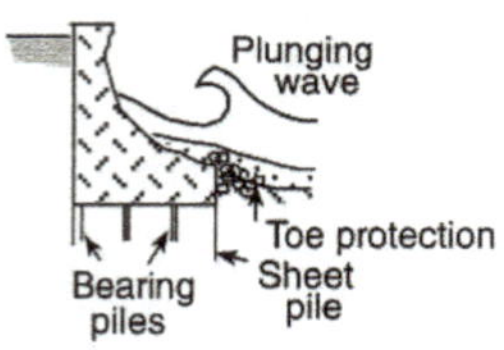	Concave structure introduces a dissipative element. The curve can prevent waves from overtopping the wall and provides extra protection for the toe of the wall Curved seawalls aim to re-direct most of the incident energy, resulting in low reflected waves and much reduced turbulence.	More complex engineering and design process. The deflected waves can scour material at the base of the wall causing them to become undermined.	
Mound	Mound type seawalls, using revetments or riprap, are used in less demanding settings where lower energy erosional processes operate. The least exposed sites involve the lowest-cost bulkheads and revetments of sand bags or geotextiles. These serve to armour the shore and minimise erosion and may be either watertight or porous, which allows water to filter through after the wave energy has been dissipated.[16]			
	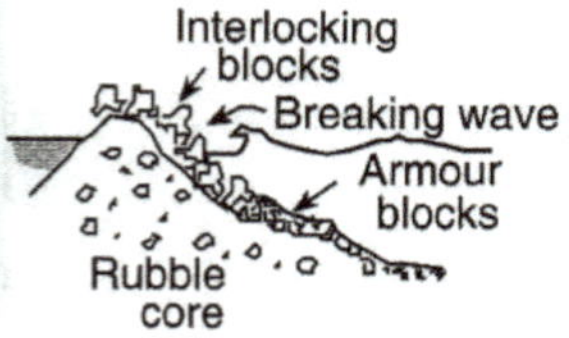	Current designs use porous designs of rock, concrete armour. Slope and loose material ensure maximum dissipation of wave energy. Lower cost option.	Less durable. Shorter life expectancy. Cannot withstand or protect from high-energy conditions effectively.	

Trade-Offs

A cost benefit approach is an effective way to determine whether a seawall is appropriate and whether the benefits are worth the expense. Besides controlling erosion, consideration must be given to the effects of hardening a shoreline on natural coastal ecosystems and human property or activities. A seawall is a static feature which can conflict with the dynamic nature of the coast and impede the exchange of sediment between land and sea. The table below summarises some positive and negative effects of seawalls which can be used when comparing their effectiveness with other coastal management options, such as beach nourishment.

Advantages and disadvantages of seawalls according to Short (1999)

Advantages	Disadvantages
Long term solution in comparison to soft beach nourishment.	Very expensive to construct.
Effectively minimizes loss of life in extreme events and damage to property caused by erosion.	Can cause beaches to dissipate rendering them useless for beach goers.
Can exist longer in high energy environments in comparison to 'soft' engineering methods.	Scars the very landscape that they are trying to save and provides an 'eyesore.'
Can be used for recreation and sightseeing.	Reflected energy of waves leading to scour at base.
Forms a hard and strong coastal defense.	Can disrupt natural shoreline processes and destroy shoreline habitats such as wetlands and intertidal beaches.
	Altered sediment transport processes can disrupt sand movement that can lead to increased erosion down drift from the structure.

Generally seawalls can be a successful way to control coastal erosion, but only if they are constructed well and out of materials which can withstand the force of ongoing wave energy. Some understanding is needed of the coastal processes and morphodynamics specific to the seawall location. Seawalls can be very helpful; they can offer a more long-term solution than soft engineering options, additionally providing recreation opportunities and protection from extreme events as well as everyday erosion. Extreme natural events expose weaknesses in the performance of seawalls, and analyses of these can lead to future improvements and reassessment.

Issues

Sea Level Rise

Sea level rise creates an issue for seawalls worldwide as it raises both the mean normal water level and the height of waves during extreme weather events, which the current seawall heights may be unable to cope with (Allan *et al.* 1999). The most recent analyses of long, good-quality tide gauge records (corrected for GIA and when possible for other vertical land motions by the Global Positioning System, GPS) indicate a mean rate of sea level rise of 1.6–1.8 mm/yr over the twentieth century. The Intergovernmental Panel on Climate Change (IPCC) (1997) suggested that sea level rise over the next 50 – 100 years will accelerate with a projected increase in global mean sea level of +18 cm by 2050 AD. This data is reinforced by Hannah (1990) who calculated similar statistics including a rise of between +16-19.3 cm throughout 1900–1988. Super storm Sandy of 2012 is an example of the devastating effects rising sea levels can cause when mixed with a perfect storm. According to Nabil Ismail, Professor of Coastal Engineering at the Maritime Academy for Science and Technology and S. Jeffress Williams, Dept. of Geology and Geophysics at The University of Hawaii at Manoa stated that Super storm Sandy sent a storm surge of 4–5 m onto New Jersey's and New York's barrier island and urban shorelines, estimated at $70 billion in damage. EGU General Assembly Conference Abstracts. Vol. 15. 2013. This problem could be overcome by further modelling and determining the extension of height and reinforcement of current seawalls which needs to occur for safety to be ensured in both situations.

Extreme Events

Extreme events also pose a problem as it is not easy for people to predict or imagine the strength of hurricane or storm induced waves compared to normal, expected wave patterns. An extreme event can dissipate hundreds of times more energy than everyday waves, and calculating structures which will stand the force of coastal storms is difficult and, often the outcome can become unaffordable. For example, Omaha Beach seawall in New Zealand was designed to prevent erosion from everyday waves only, and when a storm in 1976 carved out 10 m behind the existing seawall the whole structure was destroyed (GeoResources, 2001).

Other Issues

Some further issues include: lack of long term trend data of seawall effects due to a relatively short duration of data records; modelling limitations and comparisons of different projects and their effects being invalid or unequal due to different beach types; materials; currents; and environments (Christchurch City Council, 2009). Lack of maintenance is also a major issue with sea walls. In a post on October 21, 2013 by NBC reporter Lucas Seiler, more than 5,000 feet of seawall is crumbling in Punta Gorda, Florida. Residents of the area pay hundreds of dollars each year into a seawall repair program. The problem is that most of the seawalls are over a half century old and are being destroyed by only heavy downpours. If not kept in check, seawalls lose effectiveness and become expensive to repair.

Historical Examples

A seawall, made of rocks, in Paravur near Kollam city in India.

Seawall construction has existed since ancient times. In the 1st century BCE, Romans built a seawall / breakwater at Caesarea Maritima creating an artificial harbor (Sebastos Harbor). The construction used Pozzolana concrete which hardens in contact with sea water. Barges were constructed and filled with the concrete. They were floated into position and sunk. The resulting harbor / breakwater / sea wall is still in existence today - more than 2000 years later.

More recently, sea walls were constructed in 1623 in Canvey Island, UK, when great floods of the Thames estuary occurred, prompting the construction of protection for further events in this flood prone area (Council of Europe, 1999). Since then, seawall design has become more complex and intricate in response to an improvement in materials, technology and an understanding of how coastal processes operate. This section will outline some key case studies of seawalls in chronological order and describe how they have performed in response to tsunami or ongoing natural processes and how effective they were in these situations. Analysing the successes and shortcomings of seawalls during severe natural events allows their weaknesses to be exposed, and areas become visible for future improvement.

Pondicherry

On December 26, 2004, towering waves of the 2004 Indian Ocean earthquake tsunami crashed against India's south-eastern coastline killing thousands. However, the former French colonial enclave of Pondicherry escaped unscathed. This was primarily due to French engineers who had constructed (and maintained) a massive stone seawall during the time when the city was a French colony. This 300-year-old seawall effectively kept Pondicherry's historic center dry even though tsunami waves drove water 24 ft (7.3 m) above the normal high-tide mark.

The barrier was initially completed in 1735 and over the years, the French continued to fortify the wall, piling huge boulders along its 1.25 mi (2 km) coastline to stop erosion from the waves pounding the harbor. At its highest, the barrier running along the water's edge reaches about 27 ft (8.2 m) above sea level. The boulders, some weighing up to a ton, are weathered black and brown. The sea wall is inspected every year and whenever gaps appear or the stones sink into the sand, the government adds more boulders to keep it strong (Allsop, 2002).

The Union Territory of Pondicherry recorded some 600 deaths from the huge tsunami waves that struck India's coast after the mammoth underwater earthquake (which measured 9.0 on the moment magnitude scale) off Indonesia, but most of those killed were fishermen who lived in villages beyond the artificial barrier which reinforces the effectiveness of seawalls.

Vancouver

The Vancouver Seawall is a stone seawall constructed around the perimeter of Stanley Park in Vancouver. The seawall was constructed initially as waves created by ships passing through the First Narrows were eroding the area between Prospect Point and Brockton Point. The Vancouver Seawall also exemplifies how seawalls can be utilised and valued for recreational activities and coastal sightseeing. A pedestrian, cycling and roller blading pathway exists on the seawall and has been extended far outside the parameters of Stanley Park. Construction of the seawall began in 1917, and since then this pathway has become one of the most used features of the park by both locals and tourists and now extends 22 km in total (Belyea & Ross, 1992). The construction of the seawall also provided employment for relief workers during the Great Depression and seamen from HMCS *Discovery* on Deadman's Island who were facing punishment detail in the 1950s (Steele, 1985).

Overall, the Vancouver Seawall is a prime example of how seawalls can simultaneously provide shoreline protection and a source of recreation which enhances human enjoyment of the coastal environment. It also illustrates that although shoreline erosion is a natural process, human activities, interactions with the coast and poorly planned shoreline development projects can accelerate natural erosion rates.

Japan

At least 43 percent of Japan's 29,751 km (18,486 mi) coastline is lined with concrete seawalls or other structures designed to protect the country against high waves, typhoons or even tsunamis. During the 2011 Tōhoku earthquake and tsunami, the seawalls in most areas were overwhelmed. In Kamaishi, 4-metre (13 ft) waves surmounted the seawall — the world's largest, erected a few years ago in the city's harbour at a depth of 63 m (207 ft), a length of 2 km (1.2 mi) and a cost of $1.5 billion — and eventually submerged the city center.

The risks of dependence on seawalls was most evident in the crisis at the Fukushima Dai-ichi and Fukushima Dai-ni nuclear power plants, both located along the coast close to the earthquake zone, as the tsunami washed over walls that were supposed to protect the plants. Arguably, the additional defense provided by the seawalls presented an extra margin of time for citizens to evacuate and also stopped some of the full force of energy which would have caused the wave to climb higher in the backs of coastal valleys. In contrast, the seawalls also acted in a negative way to trap water and delay its retreat.

The failure of the world's largest seawall, which cost $1.5 billion to construct, shows that building stronger sea walls to protect larger areas would have been even less cost-effective. In the case of the ongoing crisis at the nuclear power plants, higher and stronger sea walls should have been built if power plants were to be built at that site. Fundamentally, the devastation in coastal areas and a final death toll predicted to exceed 10,000 could push Japan to redesign its seawalls or consider more effective alternative methods of coastal protection for extreme events. Such hardened coastlines can also provide a false sense of security to property owners and local residents as evident in this situation.

Seawalls along the Japanese coast have also been criticized for cutting settlements off from the sea, making beaches unusable, presenting an eyesore, disturbing wildlife, and being unnecessary.

REFERENCES

1. KETCHUM, B. H. 1972. The water's edge: critical problems of the coastal zone. In: Coastal Zone Workshop, 22 May-3 June 1972 Woods Hole, Massachusetts. Cambridge: MIT Press.
2. KAY, R. & ALDER, J. 1999. Coastal Planning and Management, London, E & FN Spon.
3. BROWN, K., TOMPKINS, E. L. & ADGER, N. 2002. Making Waves: Integrating coastal conservation and development, London, Earthscan Publications Limited.
4. CROOKS, S. & TURNER, R. K. 1999. Integrated coastal management: sustaining estuarine natural resources. Advances in Ecological Research, 29, 241-289.
5. BERKES, F. 1989. Common property resources: Ecology and community-based sustainable development, London.
6. CICIN-SAIN, B. 1993. Sustainable Development and Integrated Coastal Management. Ocean and Coastal Management, 21, 11-43.
7. WORLD, COMMISSION, ON, ENVIRONMENT, AND & DEVELOPMENT 1987. Towards Sustainable future, "Our common future". New York: Oxford University Press.
8. MASSELINK, G. & HUGHES, M. 2003. Introduction to Coastal Processes and Geomorphology, London, Hodder Arnold.
9. CICIN-SAIN, B. & KNECHT, R. 1998. Integrated coastal and ocean management: concepts and practices. Washington D.C.: Island Press.
10. IDRUS, M. R. 2009. Hard Habits to Break: Investigating Coastal Resource Utilisations and Management Systems in Sulawesi, Indonesia Doctor of Philosophy in Environmental Science, University of Canterbury.
11. THIA-ENG, C. 1993. Essential Elements of Integrated Coastal Zone Management. Ocean and Coastal Management, 21, 81-108.
12. WILLIAMS, A. & MICALLEF, A. 2009. Beach Management: Principles and Practice, London, Earthscan Publications Limited.
13. .Ioppolo G.,Saija G.,Salomone R. 2013. From coastal management to environmental management: The sustainable eco-tourism program for the mid-western coast of Sardinia (Italy). Land Use Policy (31), 460-471.
14. PEART, R. 2007. Beyond the Tide: Integrating the management of New Zealand's coasts, Auckland, Environmental Defence Society
15. Portal of the Iran ICZM
16. http://ec.europa.eu/environment/iczm/home.htm
17. Billé, R. (2008) "Integrated Coastal Zone Management: four entrenched illusions". S.A.P.I.EN.S. 1 (2)
18. Kurt Fedra. 2008. Integrated Coastal Zone Management (ICZM). Singapore: Research Publishing Services.
19. Yossef (2005)

20. Przedwojski et al. (1995)
21. Kamphuis, W J. (2010) Introduction to Coastal Engineering and Management. World Scientific Publishing Co Ltd. Singapore.
22. Shipman, B & Stojanovic, T. (2007) "Facts, Fictions, and Failures of Integrated Coastal Zone Management" in Europe Coastal Management. Vol. 35, Issue 2, p375 - 398 .
23. Allan, J C, Kirk, R M, Hemmingsen, M & Hart, D. (1999) Coastal Processes in Southern Pegasus Bay: A Review – A Report to Woodward-Clyde New Zealand Ltd. and the Christchurch City Council. Land and Water Studies Ltd. Christchurch.
24. Fletcher C H, Mullane R A & Richmond B M. (1997) "Beach loss along armored shorelines on Oahu, Hawaiian Islands" in Journal of Coastal Research. Vol. 13, No 3. P 209-215
25. Kraus, N & McDougal. (1996) The Effects of Seawalls on the Beach: Part I: An Updated Literature Review in Journal of Coastal Research. Vol. 12, No. 3.
26. Clarke, J R. 1994. Integrated Management of Coastal Zones. Fao Corporate Document Repository, USA.
27. Kajendra, R. (2011)
28. Masselink, G and Hughes, M J. (2003) Introduction to Coastal Processes and Geomorphology. Oxford University Press. New York. Ch 11.
29. NOAA. (2007) Shoreline Management: Alternatives to Hardening the Shore. Retrieved online 15 April 2011 from: http://coastalmanagement.noaa.gov/shoreline.html
30. GeoResources. (2001) Coastal management. Retrieved online 18 April 2011 from: http://www.georesources.co.uk/coastman.htm
31. Carter, Bill (1989). Coastal environments: an introduction to the physical, ecological, and cultural systems of coastlines. Boston: Academic Press. p. 50. ISBN 0-12-161856-0.
32. Matzner, Richard A. (2001). Dictionary of geophysics, astrophysics, and astronomy (PDF). Boca Raton: CRC Press. p. 81. ISBN 0-8493-2891-8.
33. Beer, Tom (1997). Environmental oceanography. Boca Raton: CRC Press. p. 44. ISBN 0-8493-8425-7. ... the reflected wave energy interacted with the incoming waves to produce standing waves known as clapotis, which promote erosion at the toe of the wall.
34. Milligan, J & O'riordan, T. (2007) "Governance for Sustainable Coastal Futures" in Coastal Management. Vol. 35, Issue 4, p 499–509.
35. Short, A. (1999) Handbook of Beach and Shoreface Morphodynamics. John Wiley and Sons Ltd. Ch 7.
36. MEDUS. (2011) Marine Engineering Division of University of Salerno. Retrieved online 10 April 2011 from: http://www.diciv.unisa.it/docenti/dentale/medus_.php (MEDUS)

37. Link text, "Sea level rise and its coastal impacts." Earth's Future (2013). Retrieved Online 21 February 2014.
38. Intergovernmental Panel on Climate Change. 2007. IPCC Fourth Assessment Report: Climate Change 2007. Retrieved Online 15 April 2011 from: www.ipcc.ch/publications_and_data/publications_and_data_reports
39. Hannah, J. (1990) "The Analysis of Mean Sea Level Data from New Zealand for the Period 1899–1988" in The Journal of Geophysical Research, 95, No. 88.
40. Link test, "Sea-Level Rise Implications for Coastal Protection from Southern Mediterranean to the USA Atlantic Coast."
41. Christchurch City Council. (2009) Study of the Effects of Sea Level Rise for Christchurch. Tonkin + Taylor, Christchurch.
42. , "More failing Seawalls in Punta Gorda Isles - NBC-2.com WBBH News for Fort Myers, Cape Coral & Naples, Florida." NBC-2.com. Nbc, 21 Oct. 2013. Web. 21 Feb. 2014.
43. Council of Europe. (1999) European code of conduct for coastal zones, Council of Europe, Strasbourg.
44. N.W.H. Allsop. (2002). Breakwaters, coastal structures and coastlines. Thomas Telford. ISBN 0-7277-3042-8.
45. Belyea, R (21 January 1992). The Stanley Park Technical Report. Vancouver: Stanley Park Task Force, prepared by Belyea, Sorensen & Associates. P 15.
46. Steele, Richard M. (1985). The Stanley Park Explorer. Vancouver: Whitecap Books. P 23–24.
47. https://www.cia.gov/library/publications/the-world-factbook/geos/ja.html
48. New York Times. (2011) Japan's Seawalls Didn't Provide Security from tsunami retrieved online 18 April 2011 from: http://www.thestar.com/news/world/article/953485--japan-s-seawalls-didn-t-provide-security-from-tsunami
49. Msubi (2011) Seawalls are no Match for Japan Tsunami. Retrieved online 8 April 2011 from: http://www.nippon-sekai.com/main/articles/great-east-japan-earthquake-of-2011/sea-walls-were-no-match-for-this-tsunami/
50. Lucy Craft (2014-03-11). "In Tsunami's Wake, Fierce Debate Over Japan's 'Great Wall'". NPR.

CHAPTER 9

Wastewater Engineering

Chapter Contents

- **Introduction**
- **Wastewater Treatment**
- **Water Pollution**

Wastewater engineering is a field of engineering focused on the management of water supplies for human and environmental health. It can include everything from sourcing clean supplies of drinking water to the management of wastewater in an environmentally friendly way. Specialists in this field typically have graduate degrees and can work on the design of treatment facilities, development of new equipment and techniques for processing, and a variety of other topics. Wastewater engineers also oversee activities at treatment plants on a day to day basis.

INTRODUCTION

Wastewater engineering is not usually its own degree course but a specialization from degrees such as civil engineering, environmental engineering or chemical engineering. Wastewater engineering deals with the transportation and cleaning of blackwater, greywater, and irrigation water. Wastewater treatment and water reclamation are areas of concern in this field.

Wastewater engineers map out topographical and geographical features of Earth to determine the best means of collection.

WASTEWATER TREATMENT

Wastewater treatment is a process used to convert wastewater - which is water no longer needed or suitable for its most recent use - into an effluent that can be either returned to the water cycle with minimal environmental issues or reused. The latter is called water reclamation and implies avoidance of disposal by use of treated wastewater effluent for various purposes. Treatment means removing impurities from water being treated; and some methods of treatment are applicable to both water and wastewater. The physical infrastructure used for wastewater treatment is called a wastewater treatment plant (WWTP).

The treatment of wastewater belongs to the overarching field of Public Works - Environmental, with the management of human waste, solid waste, sewage treatment, stormwater (drainage) management, and water treatment. By-products from wastewater treatment plants, such as screenings, grit and sewage sludge may also be treated in a wastewater treatment plant. If the wastewater is predominantly from municipal sources (households and small industries) it is called sewage and its treatment is called sewage treatment.

Clarifiers are widely used for wastewater treatment.

Disposal or Reuse

Although disposal or reuse occurs after treatment, it must be considered first. Since disposal or reuse are the objectives of wastewater treatment, disposal or reuse options are the basis for treatment decisions. Acceptable impurity concentrations may vary with the type of use or location of disposal. Transportation costs often make acceptable impurity concentrations dependent upon location of disposal, but expensive treatment requirements may encourage selection of a disposal location on the basis of impurity concentrations. Ocean disposal is subject to international treaty requirements. International treaties may also regulate disposal into rivers crossing international borders. Water bodies entirely within the jurisdiction of a single nation may be subject to regulations of multiple local governments. Acceptable impurity concentrations may vary widely among different jurisdictions for disposal of wastewater to evaporation ponds, infiltration basins, or injection wells.

Processes Used

Phase Separation

Phase separation transfers impurities into a non-aqueous phase. Phase separation may occur at intermediate points in a treatment sequence to remove solids generated during oxidation or polishing. Grease and oil may be recovered for fuel or saponification. Solids often require dewatering of sludge in a wastewater treatment plant. Disposal options for dried solids vary with the type and concentration of impurities removed from water.

Production of waste brine, however, may discourage wastewater treatment removing dissolved inorganic solids from water by methods like ion exchange, reverse osmosis, and distillation.

Primary settling tank of wastewater treatment plant in Dresden-Kaditz, Germany

Sedimentation

Solids and non-polar liquids may be removed from wastewater by gravity when density differences are sufficient to overcome dispersion by turbulence. Gravity separation of solids is the primary treatment of sewage, where the unit process is called "primary settling tanks" or "primary sedimentation tanks". It is also widely used for the treatment of other wastewaters. Solids that are heavier than water will accumulate at the bottom of quiescent settling basins. More complex clarifiers also have skimmers to simultaneously remove floating grease like soap scum and solids like feathers or wood chips. Containers like the API oil-water separator are specifically designed to separate non-polar liquids.

Filtration

Colloidal suspensions of fine solids may be removed by filtration through fine physical barriers distinguished from coarser screens or sieves by the ability to remove particles smaller than the openings through which the water passes. Other types of water filters remove impurities by chemical or biological processes described below.

Oxidation

Oxidation reduces the biochemical oxygen demand of wastewater, and may reduce the toxicity of some impurities. Secondary treatment converts some impurities to carbon dioxide, water, and biosolids. Chemical oxidation is widely used for disinfection.

Aeration tank of an activated sludge process at the wastewater treatment plant in Dresden-Kaditz, Germany

Biochemical Oxidation

Secondary treatment by biochemical oxidation of dissolved and colloidal organic compounds is widely used in sewage treatment and is applicable to some agricultural and industrial waste waters. Biological oxidation will preferentially remove organic compounds useful as a food supply for the treatment ecosystem. Concentration of some less digestible compounds may be reduced by co-metabolism. Removal efficiency is limited by the minimum food concentration required to sustain the treatment ecosystem. Removal efficiency of over 96% are readily attainable through proper process controls and effective maintenance of plant equipment.

Chemical Oxidation

Chemical oxidation may remove some persistent organic pollutants and concentrations remaining after biochemical oxidation. Disinfection by chemical oxidation kills bacteria and microbial pathogens by adding ozone, chlorine or hypochlorite to wastewater.

Polishing

Polishing refers to treatments made following the above methods. These treatments may also be used independently for some industrial wastewater. Chemical reduction or pH adjustment minimizes chemical reactivity of wastewater following chemical oxidation. Carbon filtering removes remaining contaminants and impurities by chemical absorption onto activated carbon. Filtration through sand (calcium carbonate) or fabric filters is the most common method used in municipal wastewater treatment.

Wastewater Treatment Plants

Wastewater treatment plants may be distinguished by the type of wastewater to be treated, i.e. whether it is sewage, industrial wastewater, agricultural wastewater or leachate.

Overview of the wastewater treatment plant of Antwerpen-Zuid, located in the south of the agglomeration of Antwerp (Belgium)

Sewage Treatment Plants

A typical municipal sewage treatment plant in an industrialized country may include primary treatment to remove solid material, secondary treatment to digest dissolved and suspended organic material as well as the nutrients nitrogen and phosphorus, and - sometimes but not always - disinfection to kill pathogenic bacteria. The sewage sludge that is produced in sewage treatment plants undergoes sludge treatment. Larger municipalities often include factories discharging industrial wastewater into the municipal sewer system. The term "sewage treatment plant" is nowadays often replaced with the term "wastewater treatment plant".

Tertiary Treatment

Tertiary treatment is a term applied to polishing methods used following a traditional sewage treatment sequence. Tertiary treatment is being increasingly applied in industrialized countries and most common technologies are micro filtration or synthetic membranes. After membrane filtration, the treated wastewater is nearly indistinguishable from waters of natural origin of drinking quality (without its minerals). Nitrates can be removed from wastewater by natural processes in wetlands but also via microbial denitrification. Ozone wastewater treatment is also growing in popularity, and requires the use of an ozone generator, which decontaminates the water as ozone bubbles percolate through the tank, but this treatment is energy intensive. Latest, and very promising treatment technology is the use aerobic granulation.

Industrial Wastewater Treatment Plants

Disposal of wastewaters from an industrial plant is a difficult and costly problem. Most petroleum refineries, chemical and petrochemical plants have onsite facilities to treat their wastewaters so that the pollutant concentrations in the treated wastewater comply with the local and/or national regulations regarding disposal of wastewaters into community treatment plants or into rivers, lakes or oceans. Constructed wetlands are being used in an increasing number of cases as they provided high quality and productive on-site treatment. Other industrial processes that produce a lot of waste-waters such as paper and pulp production has created environmental concern, leading to development of processes to recycle water use within plants before they have to be cleaned and disposed.

Industrial wastewater treatment plants are required where municipal sewage treatment plants are unavailable or cannot adequately treat specific industrial wastewaters. Industrial wastewater plants may reduce raw water costs by converting selected wastewaters to reclaimed water used for different purposes. Industrial wastewater treatment plants may reduce wastewater treatment charges collected by municipal sewage treatment plants by pre-treating wastewaters to reduce concentrations of pollutants measured to determine user fees.

Although economies of scale may favor use of a large municipal sewage treatment plant for disposal of small volumes of industrial wastewater, industrial wastewater treatment and disposal may be less expensive than correctly apportioned costs for larger volumes of industrial wastewater not requiring the conventional sewage treatment sequence of a small municipal sewage treatment plant.

An industrial wastewater treatment plant may include one or more of the following rather than the conventional primary, secondary, and disinfection sequence of sewage treatment:

- An API oil-water separator, for removing separate phase oil from wastewater.
- A clarifier, for removing solids from wastewater.
- A roughing filter, to reduce the biochemical oxygen demand of wastewater.
- A carbon filtration plant, to remove toxic dissolved organic compounds from wastewater.
- An advanced electrodialysis reversal (EDR) system with ion exchange membranes.

Agricultural Wastewater Treatment Plants

Agricultural wastewater treatment for continuous confined animal operations like milk and egg production may be performed in plants using mechanized treatment units similar to those described under industrial wastewater; but where land is available for ponds, settling basins and facultative lagoons may have lower operational costs for seasonal use conditions from breeding or harvest cycles.

Leachate Treatment Plants

Leachate treatment plants are used to treat leachate from landfills. Treatment options include: biological treatment, mechanical treatment by ultrafiltration, treatment with active carbon filters and reverse osmosis using disc tube module technology.

WATER POLLUTION

Water pollution is the contamination of water bodies (e.g. lakes, rivers, oceans, aquifers and groundwater). This form of environmental degradation occurs when pollutants are directly or indirectly discharged into water bodies without adequate treatment to remove harmful compounds.

Water pollution affects the entire biosphere – plants and organisms living in these bodies of water. In almost all cases the effect is damaging not only to individual species and population, but also to the natural biological communities.

Raw sewage and industrial waste in the New River as it passes from Mexicali to Calexico, California

Pollution in the Lachine Canal, Canada

Water pollution is a major global problem which requires ongoing evaluation and revision of water resource policy at all levels (international down to individual aquifers and wells). It has been suggested that water pollution is the leading worldwide cause of deaths and diseases, and that it accounts for the deaths of more than 14,000 people daily. An estimated 580 people in India die of water pollution related illness every day. About 90 percent of the water in the cities of China is polluted. As of 2007, half a billion Chinese had no access to safe drinking water. In addition to the acute problems of water pollution in developing countries, developed countries also continue to struggle with pollution problems. For example, in the most recent national report on water quality in the United States, 44 percent of assessed stream miles, 64 percent of assessed lake acres, and 30 percent of assessed bays and estuarine square miles were classified as polluted. The head of China's national development agency said in 2007 that one quarter the length of China's seven main rivers were so poisoned the water harmed the skin.

Water is typically referred to as polluted when it is impaired by anthropogenic contaminants and either does not support a human use, such as drinking water, or undergoes a marked shift in its ability to support its constituent biotic communities, such as fish. Natural phenomena such as volcanoes, algae blooms, storms, and earthquakes also cause major changes in water quality and the ecological status of water.

Categories

Although interrelated, surface water and groundwater have often been studied and managed as separate resources. Surface water seeps through the soil and becomes groundwater. Conversely, groundwater can also feed surface water sources. Sources of surface water pollution are generally grouped into two categories based on their origin.

Point Sources

Point source pollution – Shipyard – Rio de Janeiro.

Point source water pollution refers to contaminants that enter a waterway from a single, identifiable source, such as a pipe or ditch. Examples of sources in this category include discharges from a sewage treatment plant, a factory, or a city storm drain. The U.S. Clean Water Act (CWA) defines point source for regulatory enforcement purposes. The CWA definition of point source was amended in 1987 to include municipal storm sewer systems, as well as industrial storm water, such as from construction sites.

Non-Point Sources

Nonpoint source pollution refers to diffuse contamination that does not originate from a single discrete source. NPS pollution is often the cumulative effect of small amounts of contaminants gathered from a large area. A common example is the leaching out of nitrogen compounds from fertilized agricultural lands.Nutrient runoff in storm water from "sheet flow" over an agricultural field or a forest are also cited as examples of NPS pollution.

Blue drain and yellow fish symbol used by the UK Environment Agency to raise awareness of the ecological impacts of contaminating surface drainage

Contaminated storm water washed off of parking lots, roads and highways, called urban runoff, is sometimes included under the category of NPS pollution. However, because this runoff is typically channeled into storm drain systems and discharged through pipes to local surface waters, it becomes a point source.

Groundwater Pollution

Interactions between groundwater and surface water are complex. Consequently, groundwater pollution, also referred to as groundwater contamination, is not as easily classified as surface water pollution. By its very nature, groundwater aquifers are susceptible to contamination from sources that may not directly affect surface water bodies, and the distinction of point vs. non-point source may be irrelevant. A spill or ongoing release of chemical or radionuclide contaminants into soil (located away from a surface water body) may not create point or non-point source pollution but can contaminate the aquifer below, creating a toxic plume. The movement of the plume, called a plume front, may be analyzed through a hydrological transport model or groundwater model. Analysis of groundwater contamination may focus on soil characteristics and site geology, hydrogeology, hydrology, and the nature of the contaminants.

Causes

The specific contaminants leading to pollution in water include a wide spectrum of chemicals, pathogens, and physical changes such as elevated temperature and discoloration. While many of the chemicals and substances that are regulated may be naturally occurring (calcium, sodium, iron, manganese, etc.) the concentration is often the key in determining what is a natural component of water and what is a contaminant. High concentrations of naturally occurring substances can have negative impacts on aquatic flora and fauna.

Oxygen-depleting substances may be natural materials such as plant matter (e.g. leaves and grass) as well as man-made chemicals. Other natural and anthropogenic substances may cause turbidity (cloudiness) which blocks light and disrupts plant growth, and clogs the gills of some fish species. Many of the chemical substances are toxic. Pathogens can produce waterborne diseases in either human or animal hosts. Alteration of water's physical chemistry includes acidity (change in pH), electrical conductivity, temperature, and eutrophication. Eutrophication is an increase in the concentration of chemical nutrients in an ecosystem to an extent that increases in the primary productivity of the ecosystem. Depending on the degree of eutrophication, subsequent negative environmental effects such as anoxia (oxygen depletion) and severe reductions in water quality may occur, affecting fish and other animal populations.

Pathogens

Poster to teach people in South Asia about human activities leading to the pollution of water sources

A manhole cover unable to contain a sanitary sewer overflow.

Fecal sludge collected from pit latrines is dumped into a river at the Korogocho slum in Nairobi, Kenya.

Disease-causing microorganisms are referred to as pathogens. Although the vast majority of bacteria are either harmless or beneficial, a few pathogenic bacteria can cause disease. Coliform bacteria, which are not an actual cause of disease, are commonly used as a bacterial indicator of water pollution. Other microorganisms sometimes found in surface waters that have caused human health problems include:

- *Burkholderia pseudomallei*
- *Cryptosporidium parvum*
- *Giardia lamblia*

- *Salmonella*
- *Norovirus* and other viruses
- *Parasitic worms including the Schistosoma type*

High levels of pathogens may result from on-site sanitation systems (septic tanks, pit latrines) or inadequately treated sewage discharges. This can be caused by a sewage plant designed with less than secondary treatment (more typical in less-developed countries). In developed countries, older cities with aging infrastructure may have leaky sewage collection systems (pipes, pumps, valves), which can cause sanitary sewer overflows. Some cities also have combined sewers, which may discharge untreated sewage during rain storms.

Muddy river polluted by sediment.

Pathogen discharges may also be caused by poorly managed livestock operations.

Organic, Inorganic and Macroscopic Contaminants

Contaminants may include organic and inorganic substances.

A garbage collection boom in an urban-area stream in Auckland, New Zealand.

Organic water pollutants include:

- Detergents
- Disinfection by-products found in chemically disinfected drinking water, such as chloroform
- Food processing waste, which can include oxygen-demanding substances, fats and grease
- Insecticides and herbicides, a huge range of organohalides and other chemical compounds
- Petroleum hydrocarbons, including fuels (gasoline, diesel fuel, jet fuels, and fuel oil) and lubricants (motor oil), and fuel combustion byproducts, from storm water runoff
- Volatile organic compounds, such as industrial solvents, from improper storage.
- Chlorinated solvents, which are dense non-aqueous phase liquids, may fall to the bottom of reservoirs, since they don't mix well with water and are denser.
    - Polychlorinated biphenyl (PCBs)
    - Trichloroethylene
- Perchlorate
- Various chemical compounds found in personal hygiene and cosmetic products
- Drug pollution involving pharmaceutical drugs and their metabolites

Macroscopic Pollution in Parks Milwaukee, WI

Inorganic water pollutants include:

- Acidity caused by industrial discharges (especially sulfur dioxide from power plants)
- Ammonia from food processing waste
- Chemical waste as industrial by-products
- Fertilizers containing nutrients--nitrates and phosphates—which are found in storm water runoff from agriculture, as well as commercial and residential use

- Heavy metals from motor vehicles (via urban storm water runoff) and acid mine drainage
- Secretion of creosote preservative into the aquatic ecosystem
- Silt (sediment) in runoff from construction sites, logging, slash and burn practices or land clearing sites.

Macroscopic pollution – large visible items polluting the water – may be termed "floatables" in an urban storm water context, or marine debris when found on the open seas, and can include such items as:

- Trash or garbage (e.g. paper, plastic, or food waste) discarded by people on the ground, along with accidental or intentional dumping of rubbish, that are washed by rainfall into storm drains and eventually discharged into surface waters
- Nurdles, small ubiquitous waterborne plastic pellets
- Shipwrecks, large derelict ships.

The Brayton Point Power Station in Massachusetts discharges heated water to Mount Hope Bay.

Thermal Pollution

Thermal pollution is the rise or fall in the temperature of a natural body of water caused by human influence. Thermal pollution, unlike chemical pollution, results in a change in the physical properties of water. A common cause of thermal pollution is the use of water as a coolant by power plants and industrial manufacturers. Elevated water temperatures decrease oxygen levels, which can kill fish and alter food chain composition, reduce species biodiversity, and foster invasion by new thermophilic species.Urban runoff may also elevate temperature in surface waters.

Thermal pollution can also be caused by the release of very cold water from the base of reservoirs into warmer rivers.

Transport and Chemical Reactions of Water Pollutants

Most water pollutants are eventually carried by rivers into the oceans. In some areas of the world the influence can be traced one hundred miles from the mouth by studies using hydrology transport models. Advanced computer models such as SWMM or the DSSAM Model have been used in many locations worldwide to examine the fate of pollutants in aquatic systems. Indicator filter-feeding species such as copepods have also been used to study pollutant fates in the New York Bight, for example. The highest toxin loads are not directly at the mouth of the Hudson River, but 100 km (62 mi) south, since several days are required for incorporation into planktonic tissue. The Hudson discharge flows south along the coast due to the coriolis force. Further south are areas of oxygen depletion caused by chemicals using up oxygen and by algae blooms, caused by excess nutrients from algal cell death and decomposition. Fish and shellfish kills have been reported, because toxins climb the food chain after small fish consume copepods, then large fish eat smaller fish, etc. Each successive step up the food chain causes a cumulative concentration of pollutants such as heavy metals (e.g. mercury) and persistent organic pollutants such as DDT. This is known as bio-magnification, which is occasionally used interchangeably with bio-accumulation.

A polluted river draining an abandoned copper mine on Anglesey

Large gyres (vortexes) in the oceans trap floating plastic debris. The North Pacific Gyre, for example, has collected the so-called "Great Pacific Garbage Patch", which is now estimated to be one hundred times the size of Texas. Plastic debris can absorb toxic chemicals from ocean pollution, potentially poisoning any creature that eats it.Many of these long-lasting pieces wind up in the stomachs of marine birds and animals. This results in obstruction of digestive pathways, which leads to reduced appetite or even starvation.

Many chemicals undergo reactive decay or chemical change, especially over long periods of time in groundwater reservoirs. A noteworthy class of such chemicals is the chlorinated hydrocarbons such as trichloroethylene (used in industrial metal degreasing and electronics manufacturing) and tetrachloroethylene used in the dry cleaning industry. Both of these chemicals, which are carcinogens themselves, undergo partial decomposition reactions, leading to new hazardous chemicals (including dichloroethylene and vinyl chloride).

Groundwater pollution is much more difficult to abate than surface pollution because groundwater can move great distances through unseen aquifers. Non-porous aquifers such as clays partially purify water of bacteria by simple filtration (adsorption and absorption), dilution, and, in some cases, chemical reactions and biological activity; however, in some cases, the pollutants merely transform to soil contaminants. Groundwater that moves through open fractures and caverns is not filtered and can be transported as easily as surface water. In fact, this can be aggravated by the human tendency to use natural sinkholes as dumps in areas of karst topography.

There are a variety of secondary effects stemming not from the original pollutant, but a derivative condition. An example is silt-bearing surface runoff, which can inhibit the penetration of sunlight through the water column, hampering photosynthesis in aquatic plants.

Measurement

Environmental scientists preparing water autosamplers.

Water pollution may be analyzed through several broad categories of methods: physical, chemical and biological. Most involve collection of samples, followed by specialized analytical tests. Some methods may be conducted *in situ*, without sampling, such as temperature. Government agencies and research organizations have published standardized, validated analytical test methods to facilitate the comparability of results from disparate testing events.

Sampling

Sampling of water for physical or chemical testing can be done by several methods, depending on the accuracy needed and the characteristics of the contaminant. Many contamination events are sharply restricted in time, most commonly in association with rain events. For this reason "grab" samples are often inadequate for fully quantifying contaminant levels. Scientists gathering this type of data often employ auto-sampler devices that pump increments of water at either time or discharge intervals.

Sampling for biological testing involves collection of plants and/or animals from the surface water body. Depending on the type of assessment, the organisms may be identified for biosurveys (population counts) and returned to the water body, or they may be dissected for bioassays to determine toxicity.

Physical Testing

Common physical tests of water include temperature, solids concentrations (e.g., total suspended solids (TSS)) and turbidity.

Chemical Testing

Water samples may be examined using the principles of analytical chemistry. Many published test methods are available for both organic and inorganic compounds. Frequently used methods include pH, biochemical oxygen demand (BOD), chemical oxygen demand (COD), nutrients (nitrate and phosphorus compounds), metals (including copper, zinc, cadmium, lead and mercury), oil and grease, total petroleum hydrocarbons (TPH), and pesticides.

Biological Testing

Biological testing involves the use of plant, animal, and/or microbial indicators to monitor the health of an aquatic ecosystem. They are any biological species or group of species whose function, population, or status can reveal what degree of ecosystem or environmental integrity is present. One example of a group of bio-indicators are the copepods and other small water crustaceans that are present in many water bodies. Such organisms can be monitored for changes (biochemical, physiological, or behavioral) that may indicate a problem within their ecosystem.

Control of Pollution

Decisions on the type and degree of treatment and control of wastes, and the disposal and use of adequately treated wastewater, must be based on a consideration all the technical factors of each drainage basin, in order to prevent any further contamination or harm to the environment.

Sewage Treatment

Deer Island Wastewater Treatment Plant serving Boston, Massachusetts and vicinity.

In urban areas of developed countries, domestic sewage is typically treated by centralized sewage treatment plants. Well-designed and operated systems (i.e., secondary treatment or better) can remove 90 percent or more of the pollutant load in sewage. Some plants have additional systems to remove nutrients and pathogens.

Cities with sanitary sewer overflows or combined sewer overflows employ one or more engineering approaches to reduce discharges of untreated sewage, including:

- utilizing a green infrastructure approach to improve storm water management capacity throughout the system, and reduce the hydraulic overloading of the treatment plant
- repair and replacement of leaking and malfunctioning equipment
- increasing overall hydraulic capacity of the sewage collection system (often a very expensive option).

A household or business not served by a municipal treatment plant may have an individual septic tank, which pre-treats the wastewater on site and infiltrates it into the soil.

Industrial Wastewater Treatment

Dissolved air flotation system for treating industrial wastewater.

Some industrial facilities generate ordinary domestic sewage that can be treated by municipal facilities. Industries that generate wastewater with high concentrations of conventional pollutants (e.g. oil and grease), toxic pollutants (e.g. heavy metals, volatile organic compounds) or other non-conventional pollutants such as ammonia, need specialized treatment systems. Some of these facilities can install a pre-treatment system to remove the toxic components, and then send the partially treated wastewater to the municipal system. Industries generating large volumes of wastewater typically operate their own complete on-site treatment systems. Some industries have been successful at redesigning their manufacturing processes to reduce or eliminate pollutants, through a process called pollution prevention.

Heated water generated by power plants or manufacturing plants may be controlled with:

- cooling ponds, man-made bodies of water designed for cooling by evaporation, convection, and radiation
- cooling towers, which transfer waste heat to the atmosphere through evaporation and/or heat transfer
- cogeneration, a process where waste heat is recycled for domestic and/or industrial heating purposes.

Riparian buffer lining a creek in Iowa.

Agricultural Wastewater Treatment

Non point source controls. Sediment (loose soil) washed off fields is the largest source of agricultural pollution in the United States. Farmers may utilize erosion controls to reduce runoff flows and retain soil on their fields. Common techniques include contour plowing, crop mulching, crop rotation, planting perennial crops and installing riparian buffers.

Nutrients (nitrogen and phosphorus) are typically applied to farmland as commercial fertilizer, animal manure, or spraying of municipal or industrial wastewater (effluent) or sludge. Nutrients may also enter runoff from crop residues, irrigation water, wildlife, and atmospheric deposition. Farmers can develop and implement nutrient management plans to reduce excess application of nutrients and reduce the potential for nutrient pollution.

To minimize pesticide impacts, farmers may use Integrated Pest Management (IPM) techniques (which can include biological pest control) to maintain control over pests, reduce reliance on chemical pesticides, and protect water quality.

Feedlot in the United States

Point Source Wastewater Treatment

Farms with large livestock and poultry operations, such as factory farms, are called *concentrated animal feeding operations* or *feedlots* in the US and are being subject to increasing government regulation.Animal slurries are usually treated by containment in anaerobic lagoons before disposal by spray or trickle application to grassland. Constructed wetlands are sometimes used to facilitate treatment of animal wastes. Some animal slurries are treated by mixing with straw and composted at high temperature to produce a bacteriologically sterile and friable manure for soil improvement.

Erosion and Sediment Control from Construction Sites

Silt fence installed on a construction site.

Sediment from construction sites is managed by installation of:

- erosion controls, such as mulching and hydroseeding, and
- sediment controls, such as sediment basins and silt fences.

Discharge of toxic chemicals such as motor fuels and concrete washout is prevented by use of:

- spill prevention and control plans, and
- specially designed containers (e.g. for concrete washout) and structures such as overflow controls and diversion berms.

Control of Urban Runoff (Storm Water)

Effective control of urban runoff involves reducing the velocity and flow of storm water, as well as reducing pollutant discharges. Local governments use a variety of storm water management techniques to reduce the effects of urban runoff. These techniques, called best management practices (BMPs) in the U.S., may focus on water quantity control, while others focus on improving water quality, and some perform both functions.

Pollution prevention practices include low-impact development techniques, installation of green roofs and improved chemical handling (e.g. management of motor fuels & oil, fertilizers and pesticides). Runoff mitigation systems include infiltration basins, bioretention systems, constructed wetlands, retention basins and similar devices.

Retention basin for controlling urban runoff

Thermal pollution from runoff can be controlled by storm water management facilities that absorb the runoff or direct it into groundwater, such as bioretention systems and infiltration basins. Retention basins tend to be less effective at reducing temperature, as the water may be heated by the sun before being discharged to a receiving stream.[:p. 5–58]

Early history

Irrigation systems were invented five to seven thousand years ago as a means of supplying water to agriculture-based societies. Aqueducts and irrigation systems were among the first forms of wastewater engineering. As population centers became more dense, they were used to remove sewage from settlements. The Romans were among the first to demonstrate the effectiveness of the aqueduct. The Dark Ages marked a period where progress in water management came to a halt.

Modern History

As populations grew, the management of human waste became a growing concern and a public health threat. By the 1850s in London more than 400,000 tons of sewage were flushed into the Thames River each day - around 150 million tons a year. Diseases such as smallpox, diphtheria, measles, scarlet fever, typhus, cholera and typhoid were spread via the contaminated water supply. During the 19th century major cities started building sewage systems to remove human waste out of cities and into rivers.

During the 1900s the activated sludge process was invented. The activated sludge process is a form of water purification that uses bacteria to consume human feces. Chlorine is used to kill off the bacteria.

Over the centuries much has changed in the field of wastewater engineering. Advancements in microbiology, chemistry, and engineering have drastically changed the field. Today wastewater engineers work on the collection of clean water for drinking, chemically treating it, and using UV light to kill off micro-organisms. They also treat Water Pollution in wastewater (blackwater and greywater) so that this water may be made safe for use without endangering the population and environment around it.

Modern Challenges: Climate Change

Today the wastewater managers confront new challenges and the need for new technology as water levels decrease due to increasingly frequent and extended droughts. Technologies such as sonar mapping are being used in wells to determine the volume of water that they can hold. For example, the U.S. Geological Survey (USGS) and the State of New York worked together to map underground aquifers since the 1980s. Today they have thorough maps of these aquifers to assist in water management.

Wastewater treatment contributes to global warming in many ways. One of the factors that contributes to global warming is wastewater treatment facilities and their emissions of greenhouse gases. Some of those gases are carbon dioxide, methane, and nitrous oxide. These gases occur because of the decomposition of organic material from the anaerobic bacteria. These bacteria clean the leftover waste. Even if the anaerobic bacteria decomposition produces these gases, the percentage of greenhouse gases that other equipment produce is still greater than the contribution of the anaerobic bacteria. Also, the power usage from those machinery is very high. That is why many facilities are undergoing renovation to use higher levels of anaerobic bacteria compared to other types of equipment.

The effects of global warming on waste water treatment facilities vary from location to location. In the Northeast there has been increased rainfall causing treatment plants to be too small relative to the amount coming in. In the West there is a lack of rainfall and water sources are being depleted. This means the facilities will need to be updated to meet modern and future environmental changes such as the increased use of recycled and reclaimed water Climate change also affects piping and increases the rate of corrosion, adding to facility cost.

Desalination plants may be required in the future for those regions hardest hit by water scarcity. Desalination is a process of cleaning water by means of evaporation. Water is evaporated and it passes through membranes. The water is then cooled and condenses allowing it to flow either back into the main water line or out to sea. Desalination is not completely "green" since the plants require energy to run and due to the fact that the brine waste must managed to avoid ecological damage to the ocean.

Education Necessary for the Job

Formal education for wastewater engineers begins in high school with students taking classes such as chemistry, biology, physics, and higher mathematics including calculus. After high school most jobs require certification from a state agency. Those wanting to advance in the industry should pursue a civil engineering, mechanical engineering, environmental engineering, or a facilities engineering degree. Gaining experience through internships and working while in college is a common pathway toward advancement.

Education about waste treatment requires course work in systems design, machinery design principles, water chemistry, and similar coursework. Other classes may include Chemistry of Plant Processes, and various plant operations courses.

Initial employment in wastewater engineering can be obtained by those with and without advanced formal education. The California State Water Resources Control Board (SWRCB), for example, shows how individuals can advance through a progression of certifications as Waste Water Treatment Operators. The Board uses a five level classification system to classify water treatment facilities into categories I-V according to the population served and the complexity of the treatment system.

The Operator Certification requirements for water treatment operators and waste water treatment operators are described in detail by State law. To meet certification requirements, operators must submit an application to SWRCB, have the necessary work experience, meet the educational requirements, and pass an examination based on the knowledge, skill,and abilities described in the regulations. Operators are required to renew their certificates every three years. To be elibible for renewal, certified operators must complete a specified number of continuing education hours after the previous issuance of a certificate.

Wastewater engineers may advance in their careers through additional education and experience. With additional knowledge and experience one can become the manager of an entire plant. The accreditation body certifying the education for the degree and license is the Accreditation Board for Engineering and Technology (ABET). Over time some companies may require the wastewater engineer to continue their education to keep up with any changes in technology.

Obtaining one's master's degree is encouraged since many companies list it as a preference in selection.

In this field 76 percent of those employed have a bachelor's degree, 17 percent have a master's degree and three percent have a post-doctoral degree as of 2013. The average annual salary is approximately $83,360.

Job Description and Typical Tasks

Wastewater engineers use a variety of skills and must have knowledge of mechanical and environmental engineering. They are required to perform tasks and demonstrate knowledge in design, mathematics, English, construction, physics, chemistry, biology, management, and personnel. Wastewater engineers must have skills in complex problem solving, critical thinking, mathematics, active listening, judgement, reading comprehension, speaking, writing, science, and system analysis. Typical work activities include problem solving, communication with management and staff, gathering information, analyzing data, evaluating standards and complying with them, and communicating with others in the field.

Wastewater engineers perform these activities by combining their knowledge and skills to perform tasks. These tasks are to understand computer-aided design programs, and to conduct studies for the construction of facilities, water supply systems and collection systems. They may design systems for wastewater collection machinery, as well as system components. They may perform water flow analysis, then select designs and equipment based on government and industry standards.

Typical Employers

Wastewater engineers work for private companies, state and local governments, and special districts. As of 2012 there were 53,000 wastewater engineers employed in the United States.

REFERENCES

1. Steinbeck, SG. "An Abridged History of Onsite Wastewater: Early Years to Present" (PDF). REHS Onsite Water Protection Branch. Environmental Health Section, Division of Public Health, NC Department of Health and Human Services. Retrieved April 9, 2015.
2. "London's Victorian sewer system". Thames Water. Tames Water. Retrieved April 4, 2015.
3. Hardy, Anne (1984). "Water and the Search for Public Health in London in the Eighteenth and Nineteenth Centuries" (PDF). Medical History. 28: 250–82. doi:10.1017/s0025727300035936. PMC 1139446 . PMID 6390024.
4. Matsuo, Tomonori; et al. "Civil Engineering Vol. I" (PDF). Wastewater Management Engineering. Encyclodedia of Life Support Systems. Retrieved April 3, 2015.
5. "Ground Water Resource Mapping". New York State Department of Resource Conservation. State of New York. Retrieved March 23, 2015.
6. Cakir, FY (September 26, 2005). "Greenhouse gas production: A comparison between aerobic and anaerobic wastewater treatment technology". Water Research. 39 (17): 4197–4203. doi:10.1016/j.watres.2005.07.042.
7. Lattemann, Sabine; et al. (2008). "Environmental impact and impact assessment of seawater desalination" (PDF). Desalination. 220 (1): 1–15. doi:10.1016/j.desal.2007.03.009. Retrieved April 20, 2015.
8. California Environmental Protection Agency. "Operator Certification Requirements TABLE" (PDF). State Water Resources Control Board. State of California. Retrieved March 30, 2015.
9. "Wastewater Treatment Plant Classification, Operator Certification, and Contract Operator Registration" (PDF). State Water Resources Control Board. State of California. Retrieved April 20, 2015.
10. "Environmental Engineer". Science Buddies. Science Buddies, Inc. Retrieved March 21, 2015.
11. "Environmental Engineer: Career Profile". Study.com. Study.com, Mountain View, CA. Retrieved March 21, 2015.

12. "Summary Report for: 17-2081.01 - Water/Wastewater Engineers". O*net Online. U.S. Department of Labor. Retrieved March 23, 2015.
13. "Wages and Employment Trends". Onetonline.org. U.S. Department of Labor. Retrieved March 25, 2015.
14. Fairfax County Department of Human Resources. "Wastewater Plant Operations Manager Classification Specification". Fairfax County Classifications. Fairfax County, VA. Retrieved March 21, 2015.
15. Bureau of Labor Statistics. "Water and Wastewater Treatment Plant and System Operators". Occupational Outlook Handbook, 2014-15 Edition. U.S. Department of Labor. Retrieved March 22, 2015.
16. Hammer, pp.300-302
17. "Sanitation". Health topics. World Health Organization.
18. "Chapter 14: Treatment, reuse and disposal of solids and biosolids". Wastewater engineering : treatment and reuse (4th ed.). Metcalf & Eddy, Inc., McGraw Hill, USA. 2003. ISBN 0-07-112250-8.
19. Weber, pp.111-138
20. Weber, pp.139-196
21. Steel & McGhee, pp.477-521
22. Weber, pp.363-408
23. Wastewater engineering : treatment and reuse (4th ed.). Metcalf & Eddy, Inc., McGraw Hill, USA. 2003. p. 1220. ISBN 0-07-112250-8.
24. Weber, p.439
25. Wastewater engineering : treatment and reuse (4th ed.). Metcalf & Eddy, Inc., McGraw Hill, USA. 2003. p. 1138. ISBN 0-07-112250-8.
26. Wastewater engineering : treatment and reuse (4th ed.). Metcalf & Eddy, Inc., McGraw Hill, USA. 2003. ISBN 0-07-112250-8.
27. Beychok, Milton R. (1967). Aqueous Wastes from Petroleum and Petrochemical Plants (1st ed.). John Wiley & Sons. LCCN 67019834.
28. Wastewater engineering : treatment and reuse (4th ed.). Metcalf & Eddy, Inc., McGraw Hill, USA. 2003. p. 1412. ISBN 0-07-112250-8.
29. J. F. Byrd, M. D. Ehrke, J. I. Whitfield. (1984) "New Bleached Kraft Pulp Plant in Georgia: State of the Art Environmental Control" Water pollution control federation 56(4): 378–385.
30. Kemmer, pp.40-4-40-11
31. Patterson, p.180
32. Kemmer, p.41-15
33. Kemmer, p.23-11
34. Patterson, p.210
35. Reed, Middlebrooks & Crites, pp.6-8

36. Pink, Daniel H. (April 19, 2006). "Investing in Tomorrow's Liquid Gold". Yahoo. Archived from the original on April 23, 2006.
37. West, Larry (2006-03-26). "World Water Day: A Billion People Worldwide Lack Safe Drinking Water". About.com.
38. "An overview of diarrhea, symptoms, diagnosis and the costs of morbidity" (PDF). CHNRI. 2010. Archived from the original (PDF)on May 12, 2013.
39. "China says water pollution so severe that cities could lack safe supplies". Chinadaily.com.cn. June 7, 2005.
40. Kahn, Joseph; Yardley, Jim (2007-08-26). "As China Roars, Pollution Reaches Deadly Extremes". New York Times.
41. Fact Sheet: 2004 National Water Quality Inventory Report to Congress (Report). Washington, D.C.: United States Environmental Protection Agency (EPA). January 2009. EPA 841-F-08-003.
42. Wachman, Richard (2007-12-09). "Water becomes the new oil as world runs dry". The Guardian. London. Retrieved 2015-09-23.
43. United States Geological Survey (USGS), Denver, CO (1998). "Ground Water and Surface Water: A Single Resource." Circular 1139.
44. United States. Clean Water Act, section 502(14), 33 U.S.C. § 1362(14).
45. U.S. CWA section 402(p), 33 U.S.C. § 1342(p)
46. Moss, Brian (2008). "Water Pollution by Agriculture" (PDF). Phil. Trans. Royal Society B. 363: 659–666. doi:10.1098/rstb.2007.2176.
47. EPA. "Protecting Water Quality from Agricultural Runoff." Fact Sheet No. EPA-841-F-05-001. March 2005.
48. C. Michael Hogan (2010). "Water pollution.". Encyclopedia of Earth. Topic ed. Mark McGinley; ed. in chief C. Cleveland. National Council on Science and the Environment, Washington, DC.
49. USGS. Reston, VA. "A Primer on Water Quality." FS-027-01. March 2001.
50. Schueler, Thomas R. "Microbes and Urban Watersheds: Concentrations, Sources, & Pathways." Reprinted in The Practice of Watershed Protection. 2000. Center for Watershed Protection. Ellicott City, MD.
51. EPA. "Illness Related to Sewage in Water." Accessed February 20, 2009. Archived April 27, 2006, at the Wayback Machine.
52. EPA. "Report to Congress: Impacts and Control of CSOs and SSOs." August 2004. Document No. EPA-833-R-04-001.
53. G. Allen Burton, Jr., Robert Pitt (2001). Stormwater Effects Handbook: A Toolbox for Watershed Managers, Scientists, and Engineers. New York: CRC/ Lewis Publishers. ISBN 0-87371-924-7.Chapter 2.
54. Schueler, Thomas R. "Cars Are Leading Source of Metal Loads in California." Reprinted in The Practice of Watershed Protection. 2000. Center for Watershed Protection. Ellicott City, MD.

55. Goel, P.K. (2006). Water Pollution - Causes, Effects and Control. New Delhi: New Age International. p. 179. ISBN 978-81-224-1839-2.

56. Kennish, Michael J. (1992). Ecology of Estuaries: Anthropogenic Effects. Marine Science Series. Boca Raton, FL: CRC Press. pp. 415–17. ISBN 978-0-8493-8041-9.

57. Laws, Edward A. (2000). Aquatic Pollution: An Introductory Text. New York: John Wiley and Sons. p. 430. ISBN 978-0-471-34875-7.

58. Zaikab, Gwyneth Dickey (2011-03-28). "Marine microbes digest plastic". Nature. Macmillan. doi:10.1038/news.2011.191. ISSN 0028-0836.

59. For example, see Baird, Rodger B.; Clesceri, Leonore S.; Eaton, Andrew D.; et al., eds. (2012). Standard Methods for the Examination of Water and Wastewater (22nd ed.). Washington, DC: American Public Health Association. ISBN 978-0875530130. (subscription required (help)).

60. Newton, David (2008). Chemistry of the Environment. Checkmark Books. ISBN 0-8160-7747-9.

61. Karr, James R. (1981). "Assessment of biotic integrity using fish communities". Fisheries. 6: 21–27. doi:10.1577/1548-8446(1981)006<0021:AOBIUF>2.0.CO;2. ISSN 1548-8446.

62. Di Luzio, Frank C. (January 1967). "Water Pollution Control: An American Must". Journal (Water Pollution Control Federation). Water Environment Federation. 39 (1): 1–7. JSTOR 25035710.

63. "Case Study: Philadelphia, Pennsylvania". Green Infrastructure Case Studies (Report). EPA. August 2010. pp. 49–51. EPA-841-F-10-004.

64. Profile of the Fossil Fuel Electric Power Generation Industry(Report). EPA. September 1997. p. 24. EPA/310-R-97-007.

65. U.S. Natural Resources Conservation Service (NRCS). Washington, DC. "National Conservation Practice Standards."National Handbook of Conservation Practices. Accessed 2015-10-02.

66. National Management Measures to Control Nonpoint Source Pollution from Agriculture (Report). EPA. July 2003. EPA-841-B-03-004.

67. "Integrated Pest Management Principles". Pest Control and Pesticide Safety for Consumers. EPA. 2015.

68. "Animal Feeding Operations". National Pollutant Discharge Elimination System. EPA. 2016.

69. Iowa Department of Natural Resources. Des Moines, IA. "Animal Feeding Operations in Iowa." Accessed March 5, 2009.

70. Tennessee Department of Environment and Conservation. Nashville, TN (2012). "Tennessee Erosion and Sediment Control Handbook."

71. Concrete Washout (Report). Stormwater Best Management Practice. EPA. February 2012. EPA 833-F-11-006.

72. "Ch. 5: Description and Performance of Storm Water Best Management Practices". Preliminary Data Summary of Urban Storm Water Best Management Practices (Report). EPA. August 1999. EPA-821-R-99-012.

73. "Low Impact Development and Other Green Design Strategies". National Pollutant Discharge Elimination System. EPA. 2014. Archived from the original on 2015-02-19.

74. California Stormwater Quality Association. Menlo Park, CA. "Municipal BMP Handbook." 2003.

75. New Jersey Department of Environmental Protection. Trenton, NJ. "New Jersey Stormwater Best Management Practices Manual." April 2004.

Index

Y